Parris's Standard Form of Building Contract
JCT 98

Parris's Standard Form of Building Contract
JCT 98

Third Edition

David Chappell

Blackwell
Science

© David Chappell and Julia Burden 2002

Blackwell Science Ltd, a Blackwell Publishing Company
Editorial Offices:
Osney Mead, Oxford OX2 0EL, UK
  Tel: +44 (0)1865 206206
Blackwell Science, Inc., 350 Main Street, Malden, MA 02148-5018, USA
  Tel: +1 781 388 8250
Iowa State Press, a Blackwell Publishing company, 2121 State Avenue, Ames, Iowa 50014-8300, USA
  Tel: +1 515 292 0140
Blackwell Science Asia Pty, 54 University Street, Carlton, Victoria 3053, Australia
  Tel: +61 (0)3 9347 0300
Blackwell Wissenschafts Verlag,
Kurfürstendamm 57, 10707 Berlin, Germany
  Tel: +49 (0)30 32 79 060

The right of the Author to be identified as the Author of this Work has been asserted in accordance with the Copyright, Designs and Patents Act 1988.

First published as *Standard Form of Building Contract: JCT 80*
First edition published by Granada Publishing 1982
Second edition published by Collins Professional and Technical Books 1985
Third edition published by Blackwell Science Ltd 2002

A catalogue record for this title is available from the British Library

ISBN 0-632-02195-0

Library of Congress
Cataloging-in-Publication Data is available

Set in 10.5/12.5pt Palatino
by DP Photosetting, Aylesbury, Bucks
Printed and bound in Great Britain by
MPG Books Ltd. Bodmin, Cornwall

The Blackwell Publishing logo is a trade mark of Blackwell Publishing Ltd

For further information on
Blackwell Science, visit our website:
www.blackwell-science.com

Contents

Preface

It was with some trepidation that I accepted the invitation to produce the third revision of this book. The original author was the eminent construction lawyer, the late Dr John Parris. I had never worked with him; indeed, I had only met him on one occasion and that very briefly. His articles and books have a special quality and none more so than the *JCT Standard Form of Building Contract*. He was never afraid to advance an unconventional point of view and his approach to the problems inherent in this contract was never less than thought provoking.

His style is impossible to duplicate and, therefore, I have made no effort to do so. Hopefully however, the book's unique approach to the contract has been preserved in the latest edition and the reader will still find throughout, sections devoted to the treatment of specific problems. In view of the length of time which has elapsed since the last edition, it has been necessary to substantially rewrite it. However, where relevant, parts of the original text have been retained albeit modified in places. The original book was full of his opinions on all aspects of the contract. Where I agree with them, they remain unaltered. The innovative structure of the book has been kept with some minor adjustment.

This never was, nor is it now, a sterile clause by clause dissertation on the contract. It is arranged on a broad topic basis and it is intended to provide guidance for architects, surveyors, project managers and contractors on the meaning of the contract and the way in which it should be operated in practice. It also attempts a few insights into some of the more difficult parts of the document. The 1998 edition incorporating amendments 1, 2 and 3 is the form under consideration and reference is made to relevant case law and legislation including, of course, the Housing Grants, Construction and Regeneration Act 1996. Adjudication is dealt with under dispute resolution together with arbitration following the Arbitration Act 1996 and the Construction Industry Model Arbitration Rules. The complex performance specified work provisions and the difficult problems of nomination are tackled and reference is made where appropriate to the nominated sub-contract conditions. The first part of the book contains guidance on related matters, such as warranties, letters of intent and *quantum meruit*.

In its original version, this was one of the few books on building contracts which I read through from start to finish. Its basis on topics rather than clauses has provided the model from which I have worked when writing other contract books.

My thanks to Julia Burden for giving access to notes prepared by Dr Parris since the last edition. I hope I have made good use of them.

I believe the law to be correct as at 31 September 2001.

David Chappell
Chappell-Marshall Limited
27 Westgate
Tadcaster
North Yorkshire
LS24 9JB

John Parris

John Parris was called to the Bar in 1946, where he undertook both commercial and criminal work, a mix not unusual in those days. He defended 19 murder cases at a time when the death penalty was still in force, which placed a particularly onerous responsibility on defending counsel. Among his cases was that of *Craig and Bentley*, where he defended Christopher Craig who had shot dead a policeman whilst breaking into a warehouse. The jury found Craig and his accomplice, Derek Bentley, guilty of murder. Craig was too young to hang but Bentley was not. For many years John Parris argued for a review of the case – finally in 1998 it was referred by the Criminal Cases Review Committee to the Court of Appeal and Bentley's conviction was quashed.

Always a keen writer, after he left the Bar John Parris spent some years in the south of France writing for a living. He produced a lively series of motoring holiday guides just as that market was developing (he had been legal advisor to Horizon Holidays in the early days of package holidays). And he wrote a biography of the Risorgimento leader, Garibaldi, as well as translating for the Folio Society the journal of a Sicilian nobleman who fought with Garibaldi's 'Thousand' when they took Sicily.

He had always enjoyed teaching work and became Head of Legal Studies at what was then the College of Building & Commerce, Stoke-on-Trent, where he also taught arbitration and construction law to surveying students. This resulted in his first book for the construction industry, *Law and Practice of Arbitrations*.

At the same time he started a legal column in *Building*, and another one for architects in *Building Design*, where he successfully campaigned with Owen Luder for the abolition of ARCUK's restriction on limited liability companies and publicity for architects. In 1976 he launched *Building Law Reports* for the Builder Group – a pioneering concept at the time when no one had addressed the need for non-lawyers to be provided with construction case reports. This and his other writing and lecturing in the field helped the industry to develop a much better understanding of construction law issues.

John Parris wrote *Standard Form of Building Contract* in 1982, shortly after the launch of JCT 80. Apart from the legal heavyweight tomes by Keating and Duncan Wallace, there was little of any substance for the industry. The book was published to excellent reviews. John Uff commented: 'This is a book you will either love or hate. In the author's own inimitable style, it is stimulatingly argumentative and full of controversy and novelty.'

John Parris died in 1996.

CHAPTER ONE
BACKGROUND TO THE CONTRACT, THE DOCUMENTS AND SOME KEY PRINCIPLES

1.1 Earlier standard forms

A standard form of contract for building work in the United Kingdom came into use in the last quarter of the nineteenth century. A copy of it, printed in *Hudson's Building Cases* (3rd edition) vol. 2 at page 632, shows that it consisted of only nineteen clauses.

From 1903, it became known as 'the RIBA contract', a title that it retained until 1977 when the term 'JCT contract' was adopted. Many members of the judiciary, however, seemed unable to adapt to the new name.

From 1903, the Standard Form of Building Contract was put together by a tripartite body consisting of representatives of the RIBA, the Construction Confederation (CC) as it is now called, and the Institute of Building (IOB), as it then was. In 1931, the IOB withdrew, so that henceforth, the body was a 'joint' one consisting of the RIBA and the National Federation of Building Trades Employers (NFBTE), now the CC.

The Standard Form was substantially rewritten in 1939, 1963 and 1980.

In 1952, the Royal Institution of Chartered Surveyors (RICS) became involved and by the year 1963, the Joint Contracts Tribunal consisted of representatives of ten bodies in the construction industry. Bodies representing sub-contractors eventually joined.

In 1980, a resolution before the RIBA Council that the Institute should withdraw entirely from the Joint Contracts Tribunal, which had drafted a new version of the Standard Form, was defeated by only one vote. Disquiet had been caused by a JCT draft of another form, the Standard Form With Contractor's Design, which made no reference whatsoever to the employment of an architect. The version finally adopted by the JCT, with which this book is not concerned, creates no obligation on either the employer or the contractor to employ an architect to produce the design, inspect the work or certify sums due to the contractor. In fact the whole philosophy of the form is quite different to the traditional form.

In any event, architects have become reconciled to the 'Design and Build' form (WCD 98), as it is now known, and it appears to have led to the employment of more architects rather than fewer, although in somewhat different, and possibly changing, roles than before. A detailed exposition of the form can be found in *The JCT Design and Build Contract* (1999) 2nd edition, by David Chappell and Vincent Powell-Smith.

1.2   *The present Joint Contracts Tribunal*

The present body responsible for drafting the current form, therefore, is broadly representative of public and private sector employers, architects, contractors and sub-contractors. The constituent bodies are:

- Royal Institute of British Architects
- Construction Confederation
- Royal Institution of Chartered Surveyors
- Local Government Association
- National Specialist Contractors Council Limited
- Association of Consulting Engineers
- Scottish Building Contract Committee
- British Property Federation.

In the late 1990s, the JCT became the Joint Contracts Tribunal Limited with a revised constitution incorporated in March 1998 and active from May 1998.

The JCT, therefore, is no longer a joint body and it has never been a tribunal. There is a standing drafting committee responsible for the actual wording of the contracts; the drafts of this committee are circulated to members and through them to their constituent bodies. It used to be the case that no wording was adopted unless the agreement of members was unanimous. The situation is slightly different since the latest review.

The constituent bodies are members of the company, each of whom nominates a director to the company board under a chairman. The JCT council, its working parties, sub-committees and the council's five colleges carry out the discussions and agreement of contract documents. When a form has reached the stage when it could be published, the directors are normally bound to publish if the council recommends it, stating that it is approved by the relevant colleges. The relevant colleges are those whose members would be a signatory to the form or to the form to which an amendment is to be issued. The relevant colleges for JCT forms in existence at May 1998 are stated in the council's standing orders.

1.3   The range of contracts

The JCT has issued a whole range of other standard forms for almost every sort of building activity including cost plus and management fee. In addition, it is understood at the time of writing that, in response to the plea in the *Latham Report* for a suite of contracts, the JCT has plans to publish a range of professional contracts which will sit happily with the building contracts. A promised Standard Construction Management Contract is still awaited, as are a full range of sub-contracts.

This book deals solely with the Standard Form of Building Contract, which will be referred to as JCT 98. Between 1980, when the contract was substantially redrafted, and the end of 1998, eighteen amendments to the form had been produced by JCT. The eighteenth, which was issued to coincide with the coming into force on 1 May 1998 of the Housing Grants, Construction and Regeneration Act 1996, took account of the Act and many of the recommendations of the *Latham Report*. It also contained errors. The form was reprinted, to take account of all the amendments and to correct the errors, at the end of 1998. It is a matter of some regret that the opportunity was not taken to make clear at the back of the form, the clauses which had been changed. Many contracts no doubt were tendered on the basis of JCT 80 with amendments including number 18 and subsequent contracts were executed on the revised JCT 98 in the mistaken belief that it was identical to JCT 80 with all amendments.

JCT 98 comes in six variants. It can be obtained With Quantities, Without Quantities or With Approximate Quantities either in a Private Edition or a Local Authorities Edition. The With Quantities variant is the one intended for major work.

Traditionally the Local Authorities Edition is regarded as the basic one for comment. It has substantially the same wording as the Private Edition but there are some omissions. In view of this and the perceived currently greater use of the Private Edition, that is the form on which this book will concentrate, noting the differences with the Local Authorities Edition where appropriate.

Readers will find the two other lump sum contracts dealt with in *The JCT Intermediate Form of Contract* (1999) 2nd edition and the *JCT Minor Works Form of Contract* (1998) 2nd edition, both by David Chappell and Vincent Powell-Smith.

Other documents used with JCT 98 are:

- Sectional Completion Supplement
- Contractor's Designed Portion Supplement
- Nominated sub-contract documents:
  - NSC/T   Part (1) Invitation to tender
            Part (2) Tender
            Part (3) Particular conditions
  - NSC/A   Agreement
  - NSC/C   Conditions
  - NSC/W   Warranty
  - NSC/N   Nomination
- Nominated supplier documents:
  - TNS/1   Tender
  - TNS/2   Warranty
- Domestic sub-contract DOM/1

1.4   *The nature of a standard form*

Standard forms of contract are like standard suits, standard cars, standard housing or standard sheets of paper. They are good enough across a broad spectrum of applications, but they are seldom entirely appropriate.

A specially made suit or a customised car is designed to match precisely the requirements of the purchaser. The standardised versions of anything are based on a notion of a majority requirement. This is the main disadvantage of a standard form of contract. The fact that JCT produces several different standard forms indicates the difficulty.

In theory, it is much better to have forms of building contract specially drafted to suit the detailed requirements of employers. In practice, such contracts would have their own particular disadvantages. Contractors may be loath to tender on that basis and architects would be unused to administering strange contracts. Mistakes and wrong assumptions would be made as the parties began to understand how each new contract worked. Just as they all got used to it, the project would be complete and the next project would have a different bespoke contract. The situation would not be quite as bad as that of course, because every contract would have certain things in common. Out of the common elements, new standard forms would emerge.

Sound advice is never to amend standard forms. That is principally because it is difficult to ensure that any amendment works correctly in the context of the form as a whole and there is a real danger that the amendment of a clause is not carried through to amend all references to it and to amend anything else which depends upon that clause. Modern building contracts are complex documents with a multitude of interlocking provisions. Amendments are often necessary to 'customise' a standard form, but they should be carried out only by specialists.

Another danger arises out of the number of standard forms. Architects and project managers commonly use a form which is not suitable for the procurement route and attendant circumstances, but whose sole advantage is that they are used to it. It is possible to drive from Birmingham to Winchester in a tank, but to go by car is infinitely preferable. Therefore, the choice of the right standard form is important.

1.5 Types of construction contracts

Construction contracts can be analysed in various ways. This is one way:

- *Fixed price contracts*
  In fixed price contracts the contractor undertakes to do the specified work for a sum not adjustable in the price of goods or labour. This is the common situation when a contractor quotes for the installation of a shower or other minor building work. It is commonly thought that if a contractor submits what he terms an 'estimate', he will not be bound by the price. Indeed, if the final price is much higher, the contractor will often remark that what he originally gave was 'just an estimate'. That is certainly the colloquial meaning and the understanding in the industry generally. However, in *Crowshaw v. Pritchard and Renwick* (1899), it was held that a contractor's estimate, depending on its terms, could amount to a firm offer so that acceptance by the employer would result in a binding contract. Mr Justice Bingham firmly rejected the suggestion that there was some custom that an estimate was not to be treated as an offer. He said: 'There is no such custom, and if there is, it is contrary to law.' On the other

hand, a 'quotation' is always an offer to do work for a specific sum which, on acceptance, becomes a binding contract.

- *Remeasurement contracts*
  In remeasurement contracts the price is based on quantities and there is an express right for the work to be remeasured after completion. The nominated sub-contract has a provision to this effect, shown by the description of the price as 'Tender Sum'. JCT 98 with approximate quantities is also a remeasurement contract as are the 'with quantities' versions in practice.

- *Lump sum contracts*
  JCT 98 is a lump sum contract in that a specific total figure is quoted, but it should be noted that the price is subject to alteration for:
  - variations
  - fluctuations in price of goods and services
  - revaluation of prime or provisional sums
  - loss and/or expense.

It is said that the only JCT contract which has ever been known to come out at the contract sum was that for the renovation of All Souls' Church in Langham Place, London, and that may justly be regarded as a miracle of divine grace. A lump sum contract may, rarely, be an 'entire contract'.

1.6   Entire contracts

Contracts where a contractor undertakes to do work for a fixed sum are, as a matter of principle, subject to the rule in *Cutter* v. *Powell* (1795): nothing is due until the whole of the work has been completed.

The second mate of a ship had contracted to sail from Jamaica to Liverpool for the sum of thirty guineas. He served for seven weeks on the voyage, but just before the ship docked at Liverpool, he died. It was held that his widow was entitled to nothing since he had not performed expressly what he had undertaken.

In fact the contract was a unilateral one whereby the defendant undertook: 'Ten days after the ship *Governor Party* arrives at Liverpool I promise to pay Mr T. Cutter the sum of thirty guineas, provided that he proceeds, continues and does his duty as second mate from here to the port of Liverpool.' From this has been derived the concept of an entire contract where one party's obligations have to be entirely fulfilled before he is entitled to any payment at all.

Similarly, in the last decade of the nineteenth century, a builder undertook to build a house for £1565 and abandoned the project halfway through. It was held that he was entitled to nothing for the work he had done: *Sumpter* v. *Hedges* (1898).

That harsh rule still applies in principle today: *Ibmac Ltd* v. *Marshall Ltd* (1968). However, it has been mitigated so far as building work is concerned by the doctrine of divisible contracts: *Hoenig* v. *Isaacs* (1952), and substantial completion: *Dakin* v. *Lee* (1916).

The doctrine of divisible contracts simply means that, although on the face of it there may appear to be only one contract, in reality it can relate to distinct operations, such as completion of the foundations, brickwork to damp proof level, etc.

Dakin v. *Lee* (1916) held that there could be 'completion' even though there were minor defects and/or some minor work still to be done. JCT 98 uses the term 'Practical Completion' and the meaning is discussed later **[12.1]**.

Had the roles been reversed in the *Sumpter* case, so that it was the employer who had refused to let the contractor complete, the contractor would have had the option of charging for the work on a *quantum meruit* basis or of claiming damages which would have included all the profit he would have made on the job.

Contractors frequently believe that even if they do not comply exactly with the contract, if they confer some or a similar benefit on the employer, they are entitled to be paid. They fall into the same error as did a shipyard which did repairs to a ship under a lump sum contract in a case called *The Liddlesdale* (1900). There, the shipyard did not comply with the contract specifications, but instead did work which was more expensive and used materials which were more suitable. It was held that they could recover nothing. English contract law is based upon promise, not on benefit conferred, and equitable restitution is applicable only where the defendant has an option to accept or reject. 'If a man, unsolicited, cleans my shoes, what can I do but put them on?' remarked a judge in an old case.

Of course, it is possible for any contract to be varied with the consent of the other party or for any departure from the specification to be ratified subsequently; but in the absence of either, a contractor is entitled to nothing even though he may have expended considerable sums and enriched the owner.

1.7 A judicial summary of the law

In *Holland Hannen & Cubitts (Northern) Ltd* v. *Welsh Health Technical Services Organisation* (1981), Judge John Newey QC, sitting as an

Official Referee, provided an admirable summary of the law relating to the JCT 63 contract. It is equally applicable to JCT 98:

'(1)   An entire contract is one in which what is described as "complete performance" by one party is a condition precedent to the liability of the other party: *Cutter* v. *Powell* (1795); *Munro* v. *Butt* (1858).

(2)   Whether a contract is an entire one is a matter of construction; it depends upon what the parties agreed. A lump sum contract is not necessarily an entire contract. For example, a contract providing for interim payments as work proceeds but for retention money to be held until completion is usually entire as to the retention moneys, but not necessarily the interim payments: Lord Justice Denning in *Hoenig* v. *Isaacs* (1952).

(3)   The test of complete performance for the purpose of an entire contract is in fact "substantial performance": *H. Dakin & Co Ltd* v. *Lee* (1916) and *Hoenig* v. *Isaacs* (1952).

(4)   What is "substantial" is not to be determined on a comparison of cost of work done and work omitted or done badly: *Kiely & Sons Ltd* v. *Medcraft* (1965); *Bolton* v. *Mahadeva* (1972).

(5)   If a party abandons performance of the contract, he cannot recover payment for work which he has completed: *Sumpter* v. *Hedges* (1898).

(6)   If a party has done something different from that which he contracted to perform then, however valuable his work, he cannot claim to have performed substantially: *Forman & Co Proprietary* v. *The Ship "Liddlesdale"* (1900).

(7)   If a party is prevented from performing his contract by default of the other party, he is excused from performance and may recover damages: dicta, by Mr Justice Blackburn in *Appleby* v. *Myers* (1867); *Mackay* v. *Dick* (1881).

(8)   Parties may agree that, in return for one party performing certain obligations, the other will pay to him a *quantum meruit* **[1.21]**.

(9)   A contract for a payment of a *quantum meruit* may be made in the same way as any other type of contract, including conduct.

(10)   A contract for a *quantum meruit* will not readily be inferred from the actions of a landowner in using something which has become physically attached to his land: *Munro* v. *Butt* (1858).

(11)   There may be circumstances in which, even though a special

contract has not been performed there may arise a new or substituted contract: it is a matter of evidence: *Whitaker* v. *Dunn* (1887).'

1.8 Bills of quantities

The most commonly used version of JCT 98 is for use with bills of quantities. This is a peculiarly British practice. Bills were described in the Simon Report of 1944, *The Placing and Management of Building Contracts* as 'putting into words every obligation or service which will be required in carrying out the building project'.

British contractors are dedicated to it, so much so that the National Federation of Building Trades Employers (now the Construction Confederation) at one time had an agreement that no member would tender for work exceeding £8000 in value without bills of quantities being used. The agreement was held to be contrary to public interest and a violation of the Restrictive Trade Practices Act 1956: *In Re Birmingham Association of Building Trades Employers' Agreement* (1963).

Where bills of quantities are used, before inviting tenders it is necessary for the architect to prepare his drawings in sufficient detail to enable a quantity surveyor to measure from them the actual amounts to be executed, sub-divided into various trades. This bill of quantities normally starts with what are termed the 'preliminaries', which are items which relate to the project as a whole, for example, the provision of site accommodation, followed by the itemised bills.

Contractors are invited to tender on the basis of the bills of quantities and to insert the total price they require. Subsequently, before the tender can be accepted, the prospective contractor must break his total price down into a rate and price for each item of the work.

In theory, this is a system which removes the necessity for each tenderer to work out for himself, as contractors in the United States and Canada have to do, the quantities of material and labour required. It should ensure that each contractor tenders on exactly the same basis. The risk of several contractors, each effectively tendering on different, perhaps wrongly measured, amounts of work is removed. In practice it does not work quite in that way.

In the first place, the architect's drawings are rarely in sufficient detail to enable a bill of quantities to be prepared with total accuracy. Often, the quantity surveyor will be left to guess what the architect may be intending and to measure something to cover the situation. The result is that any further detailed information in the form of drawings, schedules and the like will be treated under the contract as architect's instructions requiring variations which may well lead to additional costs to the employer.

Secondly, the art of evaluating from drawings the exact amount of materials and work required varies from the difficult, as in the case of an air conditioned computer room, to the impossible, as in the case of excavations.

The bills of quantities also purport to determine the way in which the inevitable variations are to be valued, to enable a fair valuation of work done to be made for the purposes of interim payment certificates and a complete revaluation of the final contract price for the final certificate. It is often thought that the final certificate is nothing more than the sum total of interim certificates issued plus any uncertified work. This is not what the contract provides [**12.7**].

Theory and practice seldom coincide. At one time contractors made sure that, in pricing, items in the bills of quantities scheduled for early construction carried most of the value of the Works. The practice was known as 'front loading'. The idea was to transfer as much of the employer's money to the contractor as soon as possible. This could result in later items being executed at a loss and desperate efforts being made to fabricate claims.

The opposite approach in times of high inflation was to 'back load' the tender in order to get the advantage of the fluctuations clause, particularly if the contractor could afford to buy in materials early.

A contractor may gamble by putting a high rate on an item of which there is a small quantity or a low rate on items of which there is a large quantity, hoping that the quantities will increase or decrease respectively. The former will net him additional profit while the latter may secure him the contract. The practice of adjusting the rates in this and other ways has been dubbed part of the contractor's commercial strategy: *Convent Hospital* v. *Eberlin & Partners* (1988).

There is a version of JCT 98 without quantities. The general view in the industry is that contractors will load their tenders when quantities are not used so that uneconomic prices will be quoted. That may not necessarily be so. MW 98 is designed for use without quantities and contractors appear happy to tender on that basis. IFC 98 can also be used with or without bills of quantities.

1.9 The Standard Method of Measurement

In order that there might be some standardisation in the way in which bills of quantities were prepared, the RICS and what was then the NFBTE (now the CC) prepared what is termed the 'Standard Method of Measurement' currently in its seventh edition – SMM7. There are, after all, no fewer than six ways in which a hole in the ground can be measured, as became apparent in the case of *Farr* v. *Ministry of Transport* (1960) on the Institution of Civil Engineers' contract, the ICE form.

JCT 98 in clause 14.1 reads:

'The quality and quantity of the work included in the Contract Sum shall be deemed to be that which is set out in the Contract Bills.'

'Deemed' means that circumstances are to be treated as existing even if manifestly they are not. Indeed, it may be contended that if something is deemed it has the effect of conceding that what is deemed is not in fact true: *Re Cosslett (Contractors) Ltd* (1997). However, clause 2.2.2.1 makes clear that the contract bills 'are to have been prepared in accordance with' SMM7. Under clause 2.2.2.2, 'any departure from the method of preparation ... or any error in description or in quantity or omission of items' must be corrected and treated as a variation.

Since it is virtually impossible to accurately reduce the minutiae of building operations into words, inevitably there must be ambiguities and gaps in any bill of quantities. The position under JCT 98 is that the contractor is likely to be entitled to payment on frequent occasions on this basis. The way clause 2.2.2.2 is expressed is sometimes said to amount to a warranty on the part of the employer that the bills of quantities are accurate. It is sometimes argued that clause 2.2.2.2 must exclude 'things that everybody must have understood are to be done but which happen to be omitted from the quantities': *Patman and Fotheringham Ltd* v. *Pilditch* (1904). However, in that case, although quantities were part of the contract, there were terms to the effect that the contractor should supply everything needed for the Works according to the true intent of the drawings, specification and quantities whether or not particularly described. Clause 2.2.2.2 of JCT 98 is the very antithesis of that.

The other case relied upon in this way is *Williams* v. *Fitzmaurice* (1858) where a contract for the construction of a house in accordance with the specification for a fixed price omitted any reference to floorboards. It was obviously an unintended omission and the court held that the contractor was not entitled to extra payment for the floorboards, because it was evident that the contractor was to do the flooring. The specification stated that 'the whole of the materials mentioned or otherwise in the foregoing particulars, necessary for the completion of the work, must be provided by the contractor'. There were no bills of quantities in that contract, still less bills which the employer had warranted as accurate.

1.10 JCT 98 documentation

Clause 1.3 defines the 'Contract Documents' as the contract drawings, the contract bills, the articles of agreement, the conditions and the appendix. It is not good practice for the tender form, nor indeed for any other extraneous material such as correspondence, to be included as part of the contract documents although that is frequently done. It is equally common for further documents to be incorporated by reference; that is to say, by being noted in some other contract document. When this happens, serious problems of interpretation inevitably follow.

It is essential that these documents correspond exactly with the documents on which the contractor based his tender, subject to any subsequent negotiated changes. The architect must make sure that he retains a set of originals of the drawings for this purpose, because inevitably they will be altered in various ways by the time the formal contract is drawn up.

1.10.1 Completing the form

The form must be completed with care. This task should not be undertaken by the quantity surveyor or by the employer's solicitor. It is a job for the architect who will administer the contract terms. If it is necessary to make amendments to the clauses in the printed form, each amendment or deletion should be clearly made in the appropriate place on the form and each party should initial, preferably at the beginning and end of the amendment especially where a deletion has been carried out. Amendments should only be made by an experienced contract specialist who is able to understand the effect of each amendment on the terms as a whole.

1.10.2   Recitals and articles of agreement

The date should be left blank until the form is executed by the last of the parties. The names and addresses of the employer and the contractor must be inserted in the space provided. Where limited companies are involved, it is sensible to insert the company registration number in brackets after the company name so that there is no possible chance of confusion in cases where companies change or even exchange names.

The first recital is important. The description of the work and its location must be entered with care, because among other things it may determine whether a future variation changes the scope of the Works.

Reference to the priced activity schedule must be deleted from the second recital if the contractor is not to provide one **[5.23]**.

The third recital contains space in which the drawings must be listed or, if there are many, reference must be made to an attached, clearly identified list. It is essential that the drawings are exactly the same drawings on which the contractor tendered.

The sixth recital should be deleted if the employer is not going to provide an information release schedule **[2.11]**.

The contract sum is to be inserted in article 2 and the name of the architect must be inserted in article 3. Normally the architect will be the name of a firm. Article 3 provides space for the name of the quantity surveyor, again likely to be the name of a firm. If either architect or quantity surveyor ceases to act for any reason, the employer must nominate a replacement within 21 days. The language is depressingly convoluted, but effectively the contractor has seven days in which to object. The reason for the objection must be capable of being thought sufficient by an adjudicator, arbitrator or judge as appropriate. There is an additional proviso in the case of a replacement architect. Article 3 states expressly that he is not entitled to disregard or overrule any certificate, opinion, decision, approval or instruction given by the former architect. The proviso clearly cannot mean that certificates or instructions given by the former architect cannot be changed – even if wrong. The proviso is there so that if it is necessary for the successor architect to make changes, they will be treated as variations and the contractor will be entitled to payment accordingly. For example, if the former architect had given instructions for the construction of a detail which, in the opinion of the new architect, would lead to trouble, the new architect could issue further instructions correcting the matter and the contractor would be paid for the rectification work.

Article 6.1 must be completed with the name and address of the planning supervisor if the CDM Regulations apply. That will be in virtually every instance.

1.10.3   Attestation

There are alternative clauses so that the contract can be executed under hand or as a deed. The most important difference between them is that the Limitation Act 1980 provides for a limitation period which is usually six years for contracts under hand and twelve years where the contract is executed as a deed. The limitation period starts to run from the date at which the breach of contract occurred. The latest date from which the period would run would be the date of practical completion, this being the latest date at which the contractor could correct any breach before offering the building as completed in accordance with the contract documents: *Tameside Metropolitan Borough Council* v. *Barlows Securities Group Services Ltd* (2001). Contractors will doubtless wish to execute all contracts under hand, but employers will seek to extend the contractor's liability for as long a period as possible and consequently are well advised to see that the contract is entered into as a deed.

Before the Law of Property (Miscellaneous Provisions) Act 1989 and the Companies Act 1989 came into force, it used to be necessary to seal a document in order to make it into a deed. Although sealing is still possible, it is no longer necessary and sealing alone will not constitute a deed. All that is required in the case of a company is that the document must state on its face that it is a deed and it must be signed by two directors or a director and a company secretary. There are slightly different requirements in the case of an individual.

1.10.4   The conditions

Clause 1.3: Amend the reference to public holidays if different public holidays are applicable.

Clause 1.10: If the parties do not wish the proper law of the contract to be the law of England, appropriate amendments must be made to clause 1.10, for example to change it to the law of Northern Ireland.

Clause 5.3.1.2: This clause should be deleted if no master programme is required although that it not an advisable course to take. If this clause is deleted, the words in parenthesis in clause 5.3.2 must also be deleted.

Clause 13.4: Users should note that although this clause contains clauses entitled alternative A and alternative B, they are not alternatives in the sense that one is to be deleted. Neither alternative must be deleted.

Clause 22.2: If it is not possible to take out insurance against the risks covered by the definition of 'All Risks Insurance', either the definition in this clause should be amended or the risks which are actually to be covered should be inserted in place of the definition. See footnote [ff].

Clause 22: Delete two of clauses 22A, 22B and 22C depending on who is to insure (see Chapter 8).

Clause 22C.1: If the employer cannot fulfil the obligations in this clause, it must be amended to suit.

Clause 35.13.5.3.4: If the contractor is not a limited company, but subject to bankruptcy laws, this clause must be amended to refer to events which produce bankruptcy.

1.10.5   Appendix

It is important that the appendix is completed so as to correspond precisely with the information given to the contractor in the invitation to tender or, if that information subsequently has been varied by agreement between the parties, the varied details must be inserted.

1.11   *Notice provisions and reckoning days*

Clause 1.7 of the contract sets out the requirements for the giving or service of notices or other documents. It only applies if the contract does not expressly state the way in which service of documents is to be achieved. Therefore, it does not apply to notices given in connection with the determination procedures in clauses 27, 28 or 28A because that clause states that service is to be carried out by means of actual delivery, special or recorded delivery. In other cases, service is to be by any effective means to any agreed address. If the parties cannot agree over service and appropriate addresses, service can be achieved by addressing the document to the last known principal business address or if the addressee is a body corporate, to that body's registered office or its principal office, provided it is prepaid and sent by post.

Clause 1.8 sets out the way in which periods of days are to be reckoned in order to comply with the Housing Grants, Construction and Regeneration Act 1996. If something must be done within a certain number of days from a particular date, the period begins on the day after that date. Days which are public holidays are excluded. Public holidays are defined in clause 1.3 as 'Christmas Day, Good Friday or a day which under the Banking and Financial Dealings Act 1971 is a bank holiday'. A footnote instructs the user to amend the definition if different public holidays apply.

The law applicable to the contract is to be the law of England no matter that the nationality, residence or domicile of any of the parties is elsewhere (clause 1.10). Where a different system of law is required, this clause must be amended. Therefore, if the Works are being carried out in Northern Ireland, the parties will probably wish the applicable law to be the law of Northern Ireland. Curiously, the applicable law of two of the bonds which are now bound into the contract is stated to be the law of England and Wales and there is no specific note to amend. However, parties who amend the law of the contract will doubtless wish to amend the applicable law of the bonds also.

1.12   The employer's representative

For some unaccountable reason, a new provision has been introduced in clause 1.9 to allow the employer to appoint a representative. It is unclear, for example, why the provision should be incorporated in JCT 98, but not in IFC 98. Where the employer decides to appoint a representative, he must give a written notice to the contractor. There is a pitfall in wait for the unwary employer, because if he does no more than that, the representative so notified to the contractor will effectively be the employer's agent to exercise all the employer's functions under the contract. If the employer does not wish to give the representative carte blanche in this way, he must state any exceptions in the notice. Among the things which the employer may wish to exclude from the written notice are the paying of the contractor and the issue of determination or adjudication notices.

Although the clause does not expressly say so, the employer may withdraw the authority at any time. What is certain is that, once notified, the contractor is entitled to treat all decisions and other actions of the representative as if given by the employer until such time as he receives written notification from the employer withdrawing the authority.

A footnote advises that neither architect nor quantity surveyor should be appointed as representative, to avoid confusion over roles – quite so. One is driven to the conclusion that the employer's representative is merely a project manager by another name. The term 'project manager' is not precise in the way that 'architect' or 'surveyor' is precise: *Pride Valley Foods Ltd* v. *Hall & Partners* (2000). Indeed, project managers of various kinds are now to be found in all walks of life, particularly perhaps in the management of manufacturing production. If the employer's representative is intended to be the project manager, it is important for all parties, not least the project manager, to remember that his powers and duties do not exceed those exercisable by the employer. Crucially, the employer's representative has no power to actually manage the project; he cannot issue instructions to the contractor, run site meetings or interfere in the business of extensions of time and certificates.

1.13   *Electronic data interchange (EDI)*

Electronic data interchange may be employed if the parties so wish. They can insert an appropriate reference into appendix 1 and then clause 1.11 states that supplemental conditions for EDI apply. These conditions are bound at the back of the contract. It is not clear why anyone should want instantaneous and constant communication of this kind. The telephone and fax will deal with most emergencies and it is generally accepted that the fax is overused. Although it may be a sensible option if used with care, the advantages of hard copy correspondence and notices must not be ignored and, indeed, hard copy is essential for certain kinds of notice. Needless to say, copies must be kept in permanent form of all communications sent or received by EDI.

In broad terms, the supplementary conditions provide that the parties will enter into an EDI agreement no later than the date on which a binding contract comes into existence between the employer and the contractor. In practice, the parties will execute the building contract and the EDI agreement at the same time. Clause 2 states that dispute resolution procedures under the building contract are to apply to the EDI agreement and they will prevail over any dispute resolution procedures in the EDI agreement.

The EDI agreement cannot override or modify anything in the contract unless the provisions expressly so state. The following must always be in writing:

- Determination of the contractor's employment
- Suspension by the contractor of his obligations
- The final certificate
- Invoking dispute resolution procedures, for example, a notice to concur in the appointment of an arbitrator
- Any agreement which the parties may enter into which amends the contract, including the EDI provisions.

1.14   *The Housing Grants, Construction and Regeneration Act 1996*

Although this is one of the most significant pieces of legislation in recent years so far as construction contracts are concerned, it is quite startling how many members of the industry have no, or only an imperfect, grasp of its provisions. (In Northern Ireland legislation to the same effect is the Construction Contracts (Northern Ireland) Order 1997.) Part II of the Act is the important part so far as construction contracts are concerned. Nothing replaces a careful study of the Act, but what follows is a general survey of the principal provisions of Part II.

Part II deals with construction contracts and every construction professional should have a copy. It is only a few pages long. Included in the definition of such contracts is an agreement 'to do architectural, design, or surveying work, or ... to provide advice on building, engineering, interior or exterior decoration or on the laying out of landscape in relation to construction operations'.

'Construction operations' are defined in some detail. Broadly they are the construction, alteration, repair, etc. of buildings, structures, roadworks, docks and harbours, power lines, sewers and the like. They also include installation of fittings such as heating, electrical or air conditioning, external or internal cleaning carried out as part of construction and site clearance, tunnelling, foundations and other preparatory work and painting or decorating. Excluded are such things as drilling for natural gas, mineral extraction, manufacture of certain components, construction or demolition of plant where the primary activity is nuclear processing, effluent treatment or chemicals, construction of artistic works, sign writing and other peripheral installations. More importantly, it does not bite where one of the parties intends to take residence in the subject of the construction operations.

The provisions of the Act apply only to 'agreements in writing' and there are detailed provisions as to what that entails. Apart from the obvious, it also covers situations where there is no signature, where the parties agree orally by reference to terms which are in writing and where agreement is alleged in arbitration by one party and not denied by the other.

The Act requires that all construction contracts must include certain provisions. They are:

- *Adjudication*
  Either party must have the right to refer disputes to adjudication with the object of obtaining a decision within 28 days of referral. A party may give notice of intention to refer at any time and the referral must take place within seven days. The 28 day deadline may be extended by up to 14 days if the referring party wishes or indefinitely if both parties agree. The adjudicator may take the initiative in ascertaining the facts and the law. In other words, the adjudicator does not have to wait until one party raises a point, but can ask for evidence. The adjudicator's decision is binding until the dispute is decided by litigation, arbitration or by agreement. The parties may agree to accept the adjudicator's decision as final. The adjudicator is not to be liable for acts or omissions unless there has been bad faith.
- *Stage payments*
  A party is entitled to stage payments unless the duration of the project is less than 45 days. The parties are free to agree the intervals between payments and the amounts of such payments.
- *Date for payment*
  Every contract must contain the means of working out the amount due and the date on which it is due and must provide a final date for payment.
- *Set-off*
  Payment may not be withheld, nor money set-off, unless notice has been given particularising the amount to be withheld and the grounds. The notice must be given no later than the agreed period before final payment.
- *Suspension of performance of obligations*
  If the amount properly due has not been paid by the final date for payment and no effective notice withholding payment has been given, a party has the right, after giving a 7 day written notice, to suspend performance of obligations under the contract until payment has been made.
- *Pay when paid*
  Except in cases of insolvency, a clause making payment depen-

dent upon receipt of money from a third party is void. This is intended to outlaw the so-called 'pay-when-paid clause', but it may not be sufficient to do so. It does not take effect if the third party is insolvent.

To the extent that a construction contract does not include these provisions, the Scheme for Construction Contracts (England and Wales) 1998 comes into effect just as if the clauses contained in the Scheme were written into the contract. Most standard form construction contracts and all the RIBA terms of engagement comply with the Act and, therefore, the Scheme is not relevant where such forms or terms are used.

1.15   *Privity of contract and the Third Party Act*

Privity of contract is an old principle that only the parties to a contract can exercise rights under that contract: *Tweddle* v. *Atkinson* (1861). Conversely, a person who is not a party to a contract cannot have obligations imposed upon him by the contract even if he knows of its terms: *McGruther* v. *Pitcher* (1904); *Adler* v. *Dickson* (1954). The established law was undoubtedly that a third person who is not a party to a contract cannot take a benefit under it. As Lord Reid said in *Scruttons Ltd* v. *Midland Silicones Ltd* (1962):

> 'I find it impossible to deny the existence of the general rule that a stranger to a contract cannot, in a question with either of the contracting parties, take advantage of the provisions of the contract, even where it is clear from the contract that some provision in it was intended to benefit him.'

In that case, one who was not a party to a contract was not protected by an exemption clause in the contract.

When strictly applied, as it usually was, injustice was sometimes perceived and the courts, while continuing to affirm the basic principle, have striven to find ways around the doctrine in deserving cases: *Beswick* v. *Beswick* (1967).

The Contracts (Rights of Third Parties) Act 1999 came into force on 11 May 2000 and it applies throughout the UK. It interferes with the principle of privity of contract by giving the entitlement to third parties, who are not parties to the contract in question, to enforce certain rights under the contract. In order to apply:

- The contract must give the third party a right; *and*
- The terms must confer a benefit (unless it is clear that the parties did not intend a benefit to be conferred); *and*
- The third party must be identified in the contract. That can be by name, by class or by description. (It should be noted that the third party may not have existed at the time the contract was entered into, e.g. a newly formed limited company).

Such a right may only be enforced in accordance with the terms of the contract, and the party against whom the third party seeks to enforce the terms may use any defences and remedies available under the contract and may raise any set-off or counterclaim. In some instances, the third party may be treated as a party to an arbitration agreement in the contract.

The parties can rescind or vary the contract in order to remove the right, but not if:

- The third party has communicated his agreement to the term; *and* the parties know that the third party has relied on the term; *or*
- It was reasonably foreseeable that the third party would rely on the term and he has relied on it.

To overcome that, the Act allows parties to include a term in the contract by which they agree to rescind or vary without the consent of the third party or setting out circumstances for the third party's consent. Most usefully, parties to a contract may expressly exclude third party rights under that contract. That seems to be the simplest approach and it is the approach favoured by the Joint Contracts Tribunal which, by amendment 2, has inserted such an excluding provision as clause 1.12 in JCT 98. So, the position under the up-to-date version of JCT 98 is the same as before the Act came into force.

It is to be noted that, although JCT 98 makes frequent references to what the architect may or must do, he is not a party to that contract. Therefore, these expressions create no contractual obligation on him to do any of those things. His obligation to act stems from his terms of engagement with the employer, his client, which usually require the architect to administer the contract. The contractor, however, has no such contractual link. Therefore, he cannot bring an action against the architect for breach of contract for failure to do that which JCT 98 stipulated he should do. Neither can he join the architect in any arbitration arising out of JCT 98 despite there being an arbitration agreement in both JCT 98 and the architect's terms of engagement. That is not to say, however, that the architect cannot be sued in tort by the contractor under the *Hedley Byrne* principles **[1.16]**.

1.16   *Collateral warranties*

It was forecast that the advent of the Contracts (Rights of Third Parties) Act 1999 would put an end to the use of collateral warranties. However, in view of the opportunities to negate the operation of the Act in respect of particular contracts, it is unlikely that the demise of collateral warranties is imminent.

Strictly speaking, a collateral warranty is a contract which runs alongside another contract and is subsidiary to it. Warranties need not be in writing. Promises made by the employer to the contractor during pre-contractual negotiations may give rise to a collateral contract or warranty. The leading case is *Shanklin Pier Ltd* v. *Detel Products Ltd* (1951) where the employer contracted with a contractor to paint the pier. Detel induced the employer to specify their paint and gave assurances regarding its quality. The paint was properly applied by the contractor, but it did not live up to Detel's promises. It was held that there was a collateral contract between the parties under which the employer could recover the amount it had to spend to put matters right. However, it is usual for warranties to be in the form of specially drafted documents. Such documents have proliferated in recent years and it is common for contractors, nominated and often domestic sub-contractors and suppliers and certainly all the consultants, to be required to execute a collateral warranty in favour of the building owner, the fund providing the money for the project and/or any number of prospective tenants. It used to be the view that such an agreement was not very important because it merely stated in contractual terms the duties which everyone knew the architect owed to a third party in tort. That view is no longer tenable if indeed it ever was.

Considering the architect's conditions of engagement; if the building suffers a design fault, only the client can take action against the architect for breach of the conditions of engagement. For example, if an architect designs a house for the client, the house is sold on to a third party and a design defect then becomes apparent, the third party cannot take action against the architect under the conditions of engagement between the architect and client. At one time, the third party might have been able to overcome this kind of problem by suing in the tort of negligence if there was no contractual relationship. But the House of Lords case of *Murphy* v.

Brentwood District Council (1990) made it very difficult for a third party to successfully sue in tort for a defective building.

The decision in *Murphy* means that if an architect negligently designs a building, recovery in the tort of negligence will only be possible if the defective design causes injury or death to a person or if it causes damage to property other than the building which is the subject of the defective design. Even then, the recovery will be limited to compensating for the injury or damage to other persons or property and will not cover rectification of the original design defect. The result is that a third party can no longer rely on suing an architect in negligence except in very particular cases such as where the action can be brought under the reliance principle set out in *Hedley Byrne* v. *Heller* (1963). The courts now emphasise the difference between contracts which are concerned with achieving specific results and contain many terms relating to quality, and tort which is concerned with remedying wrongs. To take a simple example: if a contractor badly constructs a parapet wall, that is a breach of contract for which the law lays down remedies as between the contractor and the employer. If the parapet wall is so inadequate that it collapses and injures a passer-by, that may be negligence for which the passer-by has a remedy against the contractor in tort.

The situation has been confused recently by a number of legal cases which have enabled the original party to a contract to bring an action against the other party for breach of contract even though the original party has since sold on the building to a purchaser and received full value for it: *St Martins Property Corporation Ltd and St Martins Property Investments Ltd* v. *Sir Robert McAlpine & Sons Ltd and Linden Gardens Trust Ltd* v. *Lenesta Sludge Disposals Ltd, McLaughlin & Harvey plc and Ashwell Construction Company Ltd* (1992); *Darlington* v. *Wiltshier* (1995) and *Alfred McAlpine Construction Ltd* v. *Panatown Ltd* (2000). The situations where that can occur are likely to be limited, because it appears that if the original party (but no longer the owner of the property) is to be able to take action, among other things the other party must be shown to have known at the time of entering into the contract that the building was to be sold on or tenanted. Given that qualification, the exception begins to make some kind of sense.

The purpose of a duty of care agreement is to create a contractual obligation to third parties who otherwise would be unlikely to have any remedy if problems became apparent after completion. There are a great many forms of warranty in circulation, some of which have been especially drafted by solicitors with a greater or lesser experience of the construction industry generally.

Under JCT 98, it is standard practice for a proposed nominated sub-contractor to be required to enter into the JCT Standard Form of Employer/Nominated Sub-Contractor Agreement (NSC/W) which is a collateral contract **[4.8]**. This gives employer and nominated sub-contractor certain contractual rights against each other. Such collateral contracts are desirable in order to protect the employer both as regards nominated sub-contractors and nominated suppliers in three main areas:

- Design work carried out by the nominated sub-contractor
- If the main contractor is entitled to an extension of time under the main contract due to a failure by the nominated sub-contractor
- If the main contractor is entitled to loss and/or expense under the main contract due to a failure by the nominated sub-contractor.

Delay by the nominated sub-contractor, or design failure, may be costly to the employer who, under JCT terms, has no claim against the main contractor. These and other defects are remedied by the collateral contract which gives the employer direct rights against the defaulting sub-contractor and, in return, the nominated sub-contractor is given various rights against the employer, e.g. the right to receive direct payment if the contractor defaults.

1.17   Implied terms generally

Before considering in detail the express contractual obligations in JCT 98, it is necessary to consider whether there are any terms which the law will write into the contract. A term of a contract which the parties to that contract did not expressly agree either in writing or orally and which is not inconsistent with some express term and which the law holds is part of the bargain and is binding on the parties as if it were expressly incorporated into the contract, is called an implied term. Terms may be implied in various ways. Lord Wright said in *Luxor (Eastbourne) Ltd* v. *Cooper* (1941) that there were three types of implied terms. In fact many more different types can be distinguished:

- Local custom: *Brown* v. *IRC* (1964).
- By usage in a particular trade where it has invariably been the longstanding practice in a particular trade, profession or business, e.g. 'reduced brickwork' means brick 9 inches thick: *Symonds* v. *Lloyd* (1859).

- The parties' 'own dictionary' usage – where both have agreed on using language with a meaning different to the common one: *The Karen Oltmann* (1976).
- As a basis for the doctrine of frustration of contracts where there is a supervening event which prevents performance: *Davis Contractors Ltd* v. *Fareham Urban District Council* (1956). Although this theory is no longer fashionable, it is used in this sense in many reported cases.
- Statute, e.g. under the Supply of Goods and Services Act 1982 and Sale of Goods Act 1979, the Defective Premises Act 1972 and, of course, the Housing Grants, Construction and Regeneration Act 1996.
- At common law, for example, that a contractor will supply good and proper materials and will provide completed work which is constructed in a good and workmanlike manner.
- To give the contract commercial effectiveness – the doctrine of *The Moorcock* (1889).
- If a term can be said to be the presumed intention of the parties, the courts will apply the 'officious bystander test'. Put simply, the question will be whether it can confidently be said that; 'if at the time the contract was being negotiated someone had said to the parties, 'What will happen in such and such a case ?' they would have replied: 'Of course, so and so will happen; we did not trouble to say that; it is too clear.'
- If the parties have consistently, regularly and invariably on numerous occasions dealt on certain terms and conditions, future dealings will usually be held to have been conducted on the same basis.

A term will not be implied merely because the court thinks it would have been reasonable to insert it into the contract. It is clear that there can never be an implied term to give business efficacy to a contract if there is an express term dealing with the same matter: *Les Affréteurs Réunis* v. *Leopold Walford* (1919). But it is sometimes erroneously supposed that this principle applies to all implied terms. It does not apply to those terms which are to be implied by law, i.e. under statute or at common law.

Contractors' claims may be based on breach of some implied term, e.g. on the part of the employer, not to prevent completion and to do all that is necessary on his part to bring about completion of the contract.

Implied terms which were written into the Sale of Goods Act 1893 were those which existed in common law before the law relating to the sale of goods was codified into statute. But there were, until recent years, no similar terms to be implied in building work. In part, this was due to the fact that the common law did not recognise buildings on land as anything separate from the land on which they stood.

So far as dwellings are concerned, their construction is subject to the provisions of the Defective Premises Act 1972. This statute probably imposes greater obligations on developers, contractors, sub-contractors, architects or engineers than are contained in JCT 98. Section 1(1) states:

> 'Any person taking on work for or in connection with the provision of a dwelling ... owes a duty to see that the work that he takes on is done in a workmanlike manner, with proper materials ... and so as to be fit for the purpose required ...'

Houses and flats built under the NHBC scheme were once exempted from the provisions of this statute, by virtue of a statutory instrument. That has not been the case since 1975.

For dwellings which are constructed subject to the Act and under JCT 98, there may be available a curious and ambiguous defence which is contained in section 1(2) in these terms:

> 'A person who takes on any such work for another on terms that he is to do it in accordance with instructions given by or on behalf of that other shall, to the extent to which he does it properly in accordance with those instructions, be treated for the purposes of the Section as discharging the duty imposed on him by Sub-section (1) above except where he owes a duty to that other to warn him of any defects in the instructions and failed to discharge that duty.'

The Act specifies:

> 'a person shall not be treated for the purposes of Sub-section (2) above as having been given instructions for the doing of work merely because he has agreed to the work being done in a specified manner, with specified materials or to a specified design.'

This situation leaves open to argument that a contractor working under JCT 98 has no liability under the Act since he is required by clauses 2 and 4, among others, to comply with 'all instructions issued to him by the architect' in regard to matters which the architect is by the contract authorised to issue.

Therefore, it seems likely that a contractor who supplies and sells a finished product, where he is responsible for the design, materials and construction, to the instructions of a lay client will not be able to rely on this defence. But a contractor working under JCT 98 with the employer's architect and with specifications, drawings and contract documents, will be able to – provided of course, that he has complied with the architect's instructions, has given appropriate warning of any defects in those instructions and provided a reasonably competent contractor would not have perceived those defects.

Under section 12 of the Sale of Goods Act 1979, a condition is implied in all sales of chattels, that the seller has a right to sell the goods, that the buyer shall enjoy quiet possession of them and that they are free from any charge or incumbrance to a third party.

The position, so far as goods are concerned, is now governed by the Unfair Contract Terms Act 1977, which provides that against a person dealing as a consumer (which is defined), liability for breach of those implied warranties cannot be excluded or restricted by reference to any contract term. As against a person not dealing as a consumer, section 6(3) provides that the warranties can be excluded or restricted by reference to a contract term only so far as the term is reasonable. This fully applies to goods supplied to a building site for incorporation into the structure.

Apart from statutory intervention, the courts have progressively implied in all building contract terms that 'the builder will do his work in a good and workmanlike manner; that he will supply good and proper materials; and that it will be reasonably fit for the purpose required'. In *Hancock* v. *B.W. Brazier (Anerley) Ltd* (1966) a contractor had used an infill which was commonly accepted as suitable but which, through no fault of his, reacted with the chemicals in the ground so as to cause heave. He was held liable for breach of the implied warranty. In *Young and Marten Ltd* v. *McManus Childs Ltd* (1968) a roofing sub-contractor complied exactly with the employer's instructions to install 'Somerset 13' tiles, manufactured only by one maker, on the roof. He was held liable, when the tiles failed through a latent and undiscoverable defect, for breach of an implied warranty to supply good and proper materials.

In the case of *Test Valley Borough Council* v. *Greater London Council* (1979) the claimants' local authority predecessors entered into an agreement with the LCC, the respondent's predecessors, under powers conferred by the Town Development Act 1952, whereby the latter would erect and, when completed, sell to the claimants' predecessors some 600 houses in Andover. The houses were duly completed and handed over, but subsequently many substantial defects appeared.

The matter came before the court on a case stated by an arbitrator dealing with claims made in respect of 44 houses.

The main issue at the hearing was whether it was an implied term of the agreement that the respondents' predecessors would merely exercise reasonable skill and care in erecting the houses, or whether it was an implied term that the houses would be fit for human habitation. In other words: 'What standard or duty ought to be implied?' It was held by the High Court that:

> 'there were implied terms of the agreement that the respondents would not merely exercise reasonable care but would provide completed dwellings which were constructed (a) in a good and workmanlike manner; (b) of materials which were of good quality and reasonably fit for their purpose and (c) so as to be fit for human habitation.'

The question of implied terms in construction work again reached the House of Lords in *Independent Broadcasting Authority* v. *EMI and BICC* (1980), when the principles set out above were again approved.

Lord Fraser put forward two propositions in the course of that case:

> 'It is now well recognised that in a building contract for work and materials, a term is normally implied that the main contractor will accept responsibility to his employer for material provided by nominated sub-contractors ... and the principle that applied in *Young and Marten* in respect of materials ought in my opinion to be applied here in respect of the complete structure, including its design.'

1.18   *Limitation periods for breach of contract*

The Limitation Act 1980, a consolidating statute like its predecessors, specifies a limitation of six years for actions based on simple contract.

The Limitation Act does not extinguish the right to sue: it merely limits the period within which any particular claimant must commence his action if he is not to be barred by lapse of time or 'statute barred'. Since the Act does not extinguish a right of action, unless the defendant raises the point in his defence, the claimant can proceed even if it is 20 years since the work was completed.

The six year period (twelve years in the case of a contract executed as a deed) begins to run when the cause of action accrues. In the case of a breach of contract it has been held that the cause of action accrues when the breach of contract takes place, whether the claimant knew of it or not. In *Tameside Metropolitan Borough Council* v. *Barlows Securities Group Services Ltd* (1999) in part of the judgment which was not overturned by the Court of Appeal the judge held that a contractor has two separate obligations. The first is to carry out the work in accordance with the contract during the course of the contract. He held that breaches of this obligation become actionable as they occur. The second obligation is to *complete* the Works in accordance with the contract. A distinct cause of action does not occur in that respect until practical completion has been certified. Therefore, it is not fatal if no action is taken in the first instance.

Generally speaking, therefore, a contractor can be sure that six years after he has completed any particular work, no action can be brought against him for breach of contract unless the contract is executed as a deed, when the period will be twelve years.

It is clear, therefore, that there are advantages for every employer to require the main contractor to enter into a contract as a deed. If the contractor does so, he should ensure that every sub-contractor, whether nominated or domestic, does the same. There are two exceptions to the limitation position:

- The provisions of section 32 of the Limitation Act
- If the contractor has given indemnities to the employer **[8.5]**.

Section 32 of the Limitation Act 1980 provides that time will not start to run if the right of action has been concealed by the fraud of the defendant. Normally, in law 'fraud' involves moral obliquity, a deliberate intention to cheat.

But in a succession of cases relating to building contracts, the courts have placed a meaning on the word 'fraud' entirely inconsistent with its natural or normal meaning. It has been held to mean no more than that it would have been inequitable to allow the defendant to rely on the statutory defence provided by Parliament.

An interesting example of the operation of section 32 of the Limitation Act was *Gray and Others* v. *T.P. Bennett & Son and Others* (1987). The contractors claimed that they were protected by the Act.

The construction was carried out in 1962–63 to a design which involved concrete nibs at each floor level up to the tenth floor. The horizontal projections were intended to provide support for brickwork at each level. After bulges in the brickwork were noticed in 1979, all brickwork was opened up and it was discovered that 90% of the nibs had been hacked back, in some cases so severely that the steel reinforcement was sticking out. The court held that since the nibs had been hacked and since each must have taken at least half a day, deliberate steps must have been taken to conceal this from the architect and from the clerk of works. There was, therefore, deliberate concealment under section 32 of the Act and as November 1979 was the earliest date when the claimants reasonably could have discovered the defects, the action was not statute barred.

In *Sheldon and Others* v. *R.H.M. Outhwaite (Underwriting Agencies) Ltd* (1995), Mr Justice Saville held that there was nothing to suggest that the 1980 Act was designed to cut down, as opposed to clarify, the previous wording of what was now defined as 'deliberate concealment'. He held that if there was wrongdoing followed by deliberate concealment, the ordinary time limits were extended.

More recently, in *Brocklesby* v. *Armitage and Guest* (2001) followed in *Cave* v. *Robinson, Jarvis and Roff* (2001), the Court of Appeal's very strict view of the position was to the effect that deliberate commission of a breach of duty in circumstances in which it is unlikely to be discovered for some time did not require that the person committing the act should be aware that it amounted to a breach of duty. Fraudulent concealment arises if the act which gives rise to the breach of duty is deliberate and is unlikely to be discovered.

1.19   Limitation periods and indemnities

An indemnity is a contractual obligation by one party to reimburse another against loss. The general rule is that a person seeking to enforce an indemnity can do so only after the fact and the extent of his own liability have been determined and ascertained **[8.5]**. In *R.H. Green & Silley Weir Ltd* v. *British Railways Board* (1980), a reclamation and dredging company had filled in a railway embankment, the effect of which was alleged to have been that support was withdrawn from the claimant's land. The contractor had given the Board an express indemnity against any liability whatsoever as a condition of being allowed to fill in the embankment. When the Board was sued by the claimants, it sought to bring in the contractor as a third party under the indemnity and was met by the plea that such proceedings were statute barred under section 2(1) of the Limitation Act 1939. It was claimed that the Board came under liability, if at all, at the time when the work was done. Mr Justice Dillon rejected this view, in spite of the support for it to be found in the case of *Bosma* v. *Larsen* (1966), and said:

'Following the reasoning in *The Post Office* v. *Norwich Union* (1967) and of Mr Justice Swanick [in *County and District Properties*], I hold that time does not run against the British Railways Board in favour of [the proposed third party] until the Board's liability to Green has been established and ascertained.'

In *County and District Properties Ltd* v. *Jenner & Sons* (1974), Mr Justice Swanick did, however, allow the sub-contractors to be joined as third parties even though there could be no liability on them unless, and until, he found the main contractor liable to the employer.

1.20   Letter of intent

It is quite extraordinary the number of cases in which contractors start work on the basis of a 'letter of intent'; and also cases where a contract is entered into simply by reference, e.g. 'on the JCT Minor Works terms and conditions' or for architects, 'on RIBA terms'.

Incorporating terms by reference is a dangerous practice. It ignores the fact that not only may earlier versions still exist (the case of *West Faulkner Associates* v. *London Borough of Newham*, decided in 1995, was concerned with a 1963 version of the JCT contract executed between the council and the contractors in September 1987, seven years after the successor edition to JCT 63, JCT 80, appeared) but there have been so many amendments to JCT contracts that it is impossible to say with certainty which amendment applies. Reference to the 'RIBA terms' when dealing with an architect's appointment is meaningless, because, currently, there are several forms of agreement which that term could describe.

The result is that since the parties are rarely in agreement regarding which JCT or RIBA contract is referred to, there is in many cases no contract and the one who has done work will simply be entitled to a *quantum meruit* **[1.21]**.

There may be rare occasions when a letter of intent may be construed as what may be preferable to term a unilateral contract, but which judges seem to prefer to describe as 'if' contracts. Namely, if I say to a group of people 'if you dye your hair green,

I will give you £50 each' and they do so, I am obliged to pay them £50 each.

Similarly, a letter of intent may constitute a continuing offer: 'if you start this work, we will pay you suitable remuneration'. This creates no obligation on the other party to do the work and if he does it, there are no express or implied warranties as to its quality: *British Steel Corporation* v. *Cleveland Bridge Company* (1981). Damages can never be awarded for breach of contract of a unilateral contract, since there is no obligation on the person to whom the promise is made. Mr Justice Goff said in that case:

> '... there can be no hard and fast answer to the question whether a letter of intent will give rise to a binding agreement: everything must depend on the circumstances of the case.'

Hall & Tawse South Ltd v. *Ivory Gate Ltd* (1997) is a good example of the problems which can arise when projects are commenced using what one or possibly both parties thinks of as a letter of intent.

Ivory Gate engaged Hall & Tawse to carry out refurbishment and redevelopment works. It was intended that the contract should be in JCT 80 form with Contractor's Designed Portion Supplement and heavily amended clause 19. The tender provided for two stages. In view of the need to start work on site as soon as possible, Ivory Gate sent a letter of intent to Hall & Tawse agreeing to pay 'all reasonable costs properly incurred ... as the result of acting upon this letter up to the date you are notified that you will not be appointed'. The letter proceeded to explain the work required and evinced an intention to enter into a contract in a specified sum.

Agreement was not quickly reached and Ivory Gate sent a further letter of intent. What was envisaged was that work would commence, contract details would be finalised and the signed building contract would be held in escrow (a situation where the effectiveness of the contract is subject to a condition being fulfilled). Unfortunately the contract documents were never completed. The terms of the second letter of intent were quite detailed, expressing the intention to enter into a formal contract but pending that time instructing the building contract works to commence, materials to be ordered and Hall &Tawse to act on instructions issued under the terms of the building contract. Previous letters of intent were superseded and if the works did not proceed, Hall & Tawse were to be paid all reasonable costs together with a fair allowance for overheads and profits. Two copies of the letter were provided and

Hall & Tawse were to sign one copy and return it. They neither signed nor returned the copy.

The project took about nine months longer than was planned. Liability was disputed and, therefore, the money due to Hall & Tawse was also disputed. At the time of the trial, the work was nearing completion. The judge referred to the second letter of intent as a provisional contract and said that it had been made when Hall & Tawse accepted the offer contained in it by starting work on site. It enabled the contract administrator to issue any instructions provided the instructions would be valid under the terms of the contract. The judge held that no other contract had come into existence to supplant the provisional contract and the method of determining the amounts due to Hall & Tawse was to refer to the bills of quantities which were to have formed part of the contract. The machinery for valuing the work was to be found in the JCT 80 contract. Under the provisional contract, Hall & Tawse were not entitled to stop work at any time as would have been the case under a normal letter of intent.

There were two such letters issued in this instance. One was a true letter of intent, the other was actually a contract which determined the rights and duties of the parties. Although it was intended to be provisional until a permanent contract could be executed, the absence of a subsequent permanent contract turned the provisional contract into a permanent contract. A straightforward letter of intent would have entitled the contractor to walk off site at any time and, crucially, it would have entitled the contractor to remuneration on a fair commercial rate basis which might have exceeded the contract rates.

Manchester Cabins Ltd v. *Metropolitan Borough of Bury* (1997) concerned a letter of intent which was not a contract. Tenders were invited on the basis of the JCT Standard Form of Building Contract With Contractor's Design. Although Manchester Cabins submitted a tender, it did not include the contractor's proposals. Much negotiation took place and Manchester Cabins produced some drawings. Bury sent a fax which stated:

'I am pleased to inform you that the Council has accepted your tender for the above in the sum of £41,034.24, subject to the satisfactory execution of the contract documents which will be forwarded to you in due course.'

Eventually, it was confirmed that the letter was indeed authority to commence the necessary preliminary works 'subject to the satisfactory execution of the contract documents...'. Surprisingly, later on the same day that the confirmation was sent, Bury wrote to Manchester Cabins suspending work, later stating the Council's intention to withdraw from the contract. The court held that there was no concluded contract, because the words 'subject to the satisfactory execution of the contract documents' were included. Although the phrase did not always prevent a contract from coming into effect, in view of the surrounding circumstances it was clear that there was no agreement in this instance.

Starting work on the basis of a letter of intent or terms incorporated by reference are, therefore, clearly recipes for litigation. It is far better for an employer and a contractor at an early stage to enter into a formal agreement in the current JCT or other form accepted by both parties.

1.21   Quantum meruit

The words *'quantum meruit'* literally mean 'as much as is deserved'. They are often used instead of *quantum valebat* which mean 'as much as something is worth'. It is rare for a distinction to be drawn between the expressions. The words are used in four different situations:

- If work is done under a contract which has no express provision as to price
- If there is an express agreement to pay a reasonable sum
- If work is done under what was assumed to be a valid contract, but which turns out to be void
- If work is carried out by one party at the request of another, e.g. following a letter of intent **[1.20]**.

In *Gilbert & Partners* v. *Knight* (1968), building surveyors undertook to supervise alterations to the defendant's house for the sum of £30. The alterations, originally estimated at £600, finally came to £2283. The surveyors sent the defendant a bill for £135 – the £30 originally agreed and 100 guineas they reasonably thought they were entitled to for extra work. They recovered nothing. The Court of Appeal held that there had been no fresh agreement and, therefore, there were no circumstances in which a promise to pay a *quantum meruit* could be implied.

An architect cannot certify for anything other than what the JCT contract authorises. Therefore, without the express authority of the employer, he cannot certify *quantum meruit* payments. The payment on a *quantum meruit* basis is not to be taken to mean that payment will be calculated on a 'cost plus' basis with an allowance for profit, but rather on the basis of a fair commercial rate: *Laserbore* v. *Morrison Biggs Wall* (1993).

Serck Controls Ltd v. *Drake & Scull Engineering Ltd* (2000) is an instructive case. Drake & Scull had given a letter of intent to Serck instructing them to carry out certain design and installation work on a control system for BNFL. The letter of intent stated:

> 'In the event that we are unable to agree satisfactory terms and conditions in respect of the overall package, we would undertake to reimburse you with all reasonable costs involved, provided that any failure/default can reasonably be construed as being on our part.'

The question for the judge to answer concerned the way in which the *quantum meruit* was to be calculated. Several points of interest were considered. The first was whether 'reasonable sum' meant the value to Drake & Scull or Serck's reasonable costs in carrying out the work. Judge Hicks thought that *quantum meruit* covered the whole spectrum from one to the other of these positions. However, in this case, the reference to 'reasonable sums incurred' entitled Serck to reasonable remuneration. Although the word 'costs' implied the exclusion of profit and, possibly, overheads, the judge did not believe that use of the word in this instance was intended to exclude these elements.

Next, the judge considered what, if any, relevance was to be placed on the tender. He held that, because the tender did not form part of any contract, its use was limited. It could not be the starting point for the calculation of the reasonable sum. Indeed perhaps its only use was as a check to see whether the total amount arrived at by other means was surprising.

The judge was in no doubt that Serck were not obliged to comply with any programme, because there was no contract. Nevertheless, Serck could not simply ignore the presence of other contractors and was under a duty not to interfere with them and to co-operate so far as consistent with its own legitimate commercial interests. So far as site conditions were concerned, the judge said that if the criterion was the value to Drake & Scull, site conditions in carrying out the work would be irrelevant. If the starting point had been an agreed price, the only relevant points would have been any changes to the basis of the price. However, on the basis of a reasonable remuneration, the conditions under which the work was actually undertaken were clearly relevant: if the work proved to be more difficult than expected, Serck were entitled to be recompensed for that.

The judge finally considered the conduct of the two parties, especially allegations that Serck had worked inefficiently and the impact of that on the calculation of *quantum meruit*. What he said is worth repeating:

> 'If the value is being assessed on a "costs plus" basis, for example from time sheets and hourly rates for labour, then deductions should be made for time spent in repairing or repeating defective work, and for inefficient working or (as is one of the allegations here) excessive tea breaks and the like. If the value is being assessed by reference to quantities the claimant stands to gain nothing from such activities or inactivities and, if attributable to the claimant or his sub-contractors, they are irrelevant to the basic valuation; extra time and expense enter into the picture at this stage only if relied upon by the claimant as arising without fault on his part ... If such a claimant makes a claim based on extra time or expense which was in truth his own fault he should fail, but that is simply an issue of fact ...'

The judge went on to say that if there were defects remaining at completion, there should clearly be a deduction made for those, whatever the method of valuation.

CHAPTER TWO

THE ROLE OF THE ARCHITECT AND QUANTITY SURVEYOR

2.1 Duties under the JCT contract

The duties laid on an architect under JCT 98 serve two purposes: they delimit the architect's authority in relation to the contractor and they delimit the area in which the architect is acting as authorised agent of the employer.

In *Stockport Metropolitan Borough Council v. William O'Reilly* (1978), the issues before an arbitrator were: the extent to which the Works were varied; the determination of the contractor's employment and whether that was valid under clause 25 of JCT 63 or amounted to repudiation of the contract; whether the contractor had repudiated the contract. In determining whether the contractor had repudiated his obligations or not, the arbitrator failed to distinguish between acts done by the architect which were within the scope of his authority under JCT 63, and other acts and orders given to the contractor which were unauthorised by the contract and for which the employer could not be held liable.

The arbitrator had failed to distinguish between the various acts of the architect in concluding that the contractor was justified in withdrawing from the site and treating the contract as repudiated by the employer. This amounted to an error of law on the face of the award and the award was set aside by the court.

It was stressed in *Partington & Son v. Tameside Metropolitan Borough Council* (1985) that under the JCT contracts, the architect is simply the administrator of the contract and he has no power to modify or supplement the terms of the contract as agreed between the parties. He also decided that an arbitrator had no power to create new rights, obligations and liabilities in the parties.

It is not very well understood by contractors and, it must be said, by architects, that the architect has little room for the exercise of his discretion under the JCT Standard Form. As one judge memorably said: 'It is circumscribed almost to the point of extinction'. An architect recently said: 'The contract says that retention should not be released until the issue of the certificate of practical completion, but I released half the retention some weeks earlier, because I deemed it reasonable to do so'. Clearly, that architect was guilty of professional negligence and breach of his obligations to his client. Yet the myth remains that the architect can ignore the terms of the contract and act as sole arbiter in such matters. The reality is simply stated: the architect *must* do precisely what the contract requires him to do (his duties) and he *may* do certain other things provided he complies with any attached conditions (his powers). The architect's powers and duties are set out in Table 2.1.

Table 2.1
Powers and duties of the architect under JCT 98

Clause	Power/Duty
1.7	**Duty** Serve notices as specified.
2.3	**Duty** Issue instructions after being notified in writing of a discrepancy in documents.
2.4.1	**Duty** Issue instructions after being notified in writing of a discrepancy or divergence between the contractor's statement in connection with performance specified work and an architect's instruction.
3	**Duty** Take account of any amount ascertained in part or in whole when computing the next interim certificate.
4.1.1	**Power** Issue written instructions which the contract empowers him to issue.
4.1.2	**Power** Issue seven days written notice requiring compliance with an instruction.
4.2	**Duty** Specify in writing the empowering clause in response to the contractor's request.
4.3.2	**Power** Dissent in writing from the contractor's confirmation of an instruction.
4.3.2.1	**Power** Confirm, in writing within seven days, instructions issued otherwise than in writing.
4.3.2.2	**Power** Confirm, in writing at any time before the issue of the final certificate, an instruction issued otherwise than in writing.

Clause	Power/Duty
	Table 2.1 *Contd*
5.1	**Duty** Hold the contract documents for inspection by the employer or the contractor at all reasonable times.
5.2	**Duty** Provide the contractor, immediately after execution of the contract, without charge with a certified copy of the contract documents and two further copies of the contract drawings and bills of quantities.
5.3.1.1	**Duty** Provide the contractor, immediately after execution of the contract, without charge with two copies of any necessary descriptive schedules.
5.4.1	**Duty** Release two copies of the information referred to and at the time stated in the information release schedule unless prevented by the contractor's default or otherwise agreed with the contractor.
5.4.2	**Duty** Provide two copies of such further drawings or details as are reasonably necessary to explain or amplify the contract drawings and issue instructions to enable the contractor to carry out and complete the Works in accordance with the contract. The architect must have regard to the progress of the Works or, if the contractor is likely to finish before the completion date, the architect is to have regard to the completion date.
5.6	**Power** Request the contractor to return all drawings, details, etc. which bear the architect's name, after payment of the final certificate.
5.7	**Duty** Not to divulge rates and prices in the bills of quantities.
5.8	**Duty** Issue any certificate to the employer with a copy to the contractor, unless otherwise expressly provided.
6.1.3	**Duty** Issue instructions within seven days relating to any divergence between statutory requirements and the contract documents or any instruction notified by the contractor or discovered by the architect.
6.1.6	**Duty** Issue instructions relating to any divergence between statutory requirements and the contractor's statement in connection with performance specified work if the contractor has informed the architect in writing of proposals for rectifying the divergence.
7	**Duty** Determine levels and provide accurately dimensioned drawings to enable the contractor to set out the building at ground level. **Power** Instruct the contractor not to amend errors in setting out if the employer consents.

Table 2.1 *Contd*

Clause	Power/Duty
8.1.4	**Power** Consent to the substitution of materials or goods for any contained in the contractor's statement for performance specified work.
8.2.1	**Power** Request vouchers to prove that materials and goods comply with the contract.
8.2.2	**Duty** Express any dissatisfaction with materials, goods or workmanship, which are to be to the architect's reasonable satisfaction, within a reasonable time of execution.
8.3	**Power** Issue instruction for the opening up or testing of work or materials.
8.4.1	**Power** Issue instructions regarding the removal from site of work not in accordance with the contract.
8.4.2	**Power** Allow, and confirm in writing to the contractor, work not in accordance with the contract to remain. The employer must agree and the architect must consult the contractor.
8.4.3	**Power** Issue instructions requiring reasonably necessary variations resulting from defective work. The architect must consult the contractor.
8.4.4	**Power** Issue instructions to open up or test as is reasonable in all the circumstances after having had regard to the Code of Practice to the contract to establish the likelihood of similar non-compliance.
8.5	**Power** Issue instructions requiring reasonably necessary variations or otherwise resulting from the contractor's failure to carry out the work in a workmanlike manner. The architect must consult the contractor.
8.6	**Power** Issue instructions, but not unreasonably or vexatiously, requiring the exclusion from site of any person there employed.
11	**Power** Access to the Works and workshops where work is being prepared, at all reasonable times, but subject to restrictions to protect the proprietary rights in the work.
12.	**Power** Direct the clerk of works. Confirm a clerk of work's direction within two working days.
13.2.1	**Power** Issue instructions requiring a variation.
13.2.3	**Power** Instruct that the variation is to be carried out and valued under clause 13.4.1.

Table 2.1 *Contd*

Clause	Power/Duty
13.2.4	**Power** Sanction in writing any variation made by the contractor.
13.3	**Duty** Issue instructions about the expenditure of provisional sums in the bills of quantities and in nominated sub-contracts.
13.5.4	**Duty** Verify daywork vouchers if submitted not later than the end of the week following the week in which the work was carried out.
13A.3.2	**Duty** Immediately confirm the employer's acceptance of a clause 13A quotation and state in writing that the contractor must carry out the variation, the adjustment of the contract sum, any time adjustment and acceptance of any relevant clause 3.3A quotation under NSC/C.
13A.4	**Duty** Instruct that a variation is to be carried out and valued under the default procedure or instruct that the variation is not to be carried out, if the employer does not accept the clause 13A quotation.
16.1	**Power** Consent in writing to the removal from site of materials.
17.1	**Duty** Certify practical completion when in his opinion it has been achieved, the contractor has sufficiently complied with clause 6A.4 and, in the case of performance specified work, the contractor has provided as-built drawings.
17.2	**Duty** Specify a schedule of defects and deliver to the contractor not later than 14 days after the end of the defects liability period. **Power** Instruct the contractor not to make good defects if the employer consents.
17.3	**Power** Instruct the making good of defects whenever he considers it necessary during the defects liability period. Instruct the contractor not to make good the defects if the employer consents.
17.4	**Duty** Issue a certificate of completion of making good defects when the defects have been made good.
17.5	**Power** Certify that frost damage is due to injury which took place before practical completion.
18.1	**Duty** Issue to the contractor a written statement identifying the part of the Works taken into possession and stating the relevant date.
18.1.2	**Duty** Issue a certificate of making good of defects of the relevant part after making good has taken place.
19.2.2	**Power** Consent to sub-letting.

Table 2.1 *Contd*

Clause	Power/Duty
19.3.2.1	**Power** On behalf of the employer, add persons to the list of persons from which the contractor is to choose a sub-contractor.
21.1.2	**Duty** Receive from the contractor for inspection by the employer documentary evidence of insurance having been taken out and maintained.
21.2.1	**Power** Instruct the contractor to take out insurance in joint names in respect of liability for collapse, subsidence, heave, vibration, weakening or removal of support or lowering of ground water.
21.2.2	**Duty** Receive from the contractor the policy and premium receipts in respect of clause 21.2.1 insurance.
22A.2	**Duty** Receive from the contractor the policy and premium receipts in respect of clause 22A.1 insurance.
22A.3.1	**Duty** Receive from the contractor the policy and premium receipts in respect of clause 22A.1 annual insurance and, for inspection by the employer, documentary evidence of insurance having been taken out and maintained.
22D.1	**Duty** Inform the contractor that no insurance is required, if the appendix states that liquidated damages insurance may be required, or instruct the contractor to obtain a quotation. **Duty** Obtain from the employer such information as the contractor reasonably requires to obtain a quotation. **Duty** On receipt of the quotation from the contractor, instruct him whether or not the employer wishes the contractor to accept the quotation.
22FC.3.1.2	**Duty** Issue such instructions as are necessary to enable compliance to the extent that the remedial measures require a variation.
23.2	**Power** Issue instructions postponing work.
24.1	**Duty** Issue a certificate if the contractor fails to complete the Works by the completion date. **Duty** Issue a further certificate if a new completion date is fixed after the issue of a certificate under this clause.
25.3.1	**Duty** Give a written extension of time to the contractor stating the relevant events taken into account and the extent to which he has had regard to an instruction requiring the omission of work, if the cause of delay is a relevant event and the Works are likely to be delayed beyond the completion date as a result; *or*

Clause	Power/Duty
	Notify the contractor in writing that it is not fair and reasonable to give an extension of time. **Duty** The architect must act within 12 weeks of receipt of reasonably sufficient particulars or before the completion date if earlier.
25.3.2	**Power** After the first exercise of his duty, fix a completion date earlier than that previously fixed if it is fair and reasonable having regard to instructions issued after the last exercise of his duty. He may not alter the length of time adjustment for which a clause 13A quotation has been given and accepted or an acceptance of time adjustment has been made under clause 13.4.1.2.
25.3.3	**Power** After the completion date, fix a completion date earlier or later than previously fixed or confirm the previous date. **Duty** Within 12 weeks following the date of practical completion, fix a completion date earlier or later than previously fixed or confirm the previous date.
25.3.5	**Duty** Notify each nominated sub-contractor in writing of his extension of time decisions including confirmed acceptances of 13A quotations.
26.1	**Duty** Form an opinion about whether or not the contractor has or is likely to incur loss and/or expense. **Duty** From time to time thereafter, ascertain or instruct the quantity surveyor to ascertain the amount.
26.1.2	**Power** Request information to reasonably enable the architect to form an opinion.
26.1.3	**Power** Request details of loss and/or expense reasonably necessary for ascertainment.
26.3	**Duty** If and to the extent necessary for ascertainment, give written statement to the contractor of the extension of time given for relevant events 25.4.5.1, 25.4.5.2, 25.4.6, 25.4.8 and 25.4.12.
26.4.1	**Duty** Form an opinion about whether or not the nominated sub-contractor has or is likely to incur loss and/or expense after receipt of a written application of the nominated sub-contractor via the contractor. Ascertain or instruct the quantity surveyor to ascertain the amount.
26.4.2	**Duty** If and to the extent necessary for ascertainment, give written statement to the contractor, with a copy to the nominated sub-contractor, of the extension of time for which he gave consent for the relevant events in NSC/C 2.6.5.1, 2.6.5.2, 2.6.6, 2.6.8, 2.6.12 and 2.6.15.

Table 2.1 *Contd*

Table 2.1 *Contd*

Clause	Power/Duty
27.2.2	**Power** Give a 14 day notice in writing to the contractor specifying defaults prior to determination.
27.6.2.1	**Power** Act on behalf of the employer to require the contractor, except in the case of insolvency, to assign, without payment, to the employer the benefit of supply and sub-contracts.
27.6.3	**Power** Require the contractor to remove from the Works temporary buildings, plant, tools, etc.
27.6.4.2	**Power** Prepare a certificate setting out an account when the Works have been completed and all defects dealt with.
30.1.1.1	**Duty** Issue interim certificates in accordance with clause 30.1.2 on the dates noted in the appendix.
30.1.2.1	**Power** Request the quantity surveyor to prepare interim valuations when the architect considers them to be necessary.
30.6.1.1	**Power** Instruct the contractor to provide the quantity surveyor with all documents necessary for the adjustment of the contract sum.
30.6.1.2	**Duty** Ascertain any loss and/or expense unless previously ascertained or unless the architect has instructed the quantity surveyor to do so. Forthwith send a copy of the ascertainment and of the statement of adjustments to the contractor and relevant extracts to each nominated sub-contractor.
30.7	**Duty** Issue an interim certificate, not less than 28 days before the issue of the final certificate, including the finally adjusted contract sums for all nominated sub-contractors.
30.8.1	**Duty** Issue the final certificate and inform each nominated sub-contractor of the date of issue no later than two months after the latest of: • the end of the defects liability period • the issue of the certificate of completion of making good defects • the date on which the architect sent a copy of the ascertainment and statement to the contractor under clause 30.6.1.2.
34.2	**Duty** Issue instructions regarding antiquities reported by the contractor.
34.3.1	**Duty** Form an opinion and ascertain or instruct the quantity surveyor to ascertain loss and/or expense resulting from the contractor's compliance with the antiquities provisions.

Clause	Power/Duty
	Table 2.1 *Contd*
Clause	Power/Duty
35.2.1	**Power** Consent to sub-letting by the contractor if the contractor has successfully tendered for nominated sub-contract work.
35.5.2	**Power** Issue instructions removing the contractor's objection to a proposed nominated sub-contractor or cancel the nomination and issue an instruction omitting the work or nominating another sub-contractor. **Duty** Send a copy of the instruction to the sub-contractor.
35.6	**Duty** Issue an instruction nominating a sub-contractor. **Duty** Send a copy to the sub-contractor with a copy of the completed main contract appendix.
35.9	**Duty** After receipt of a clause 35.8 notice from the contractor, within a reasonable time either: • if the contractor has notified the architect the date when he expects to have complied with clause 35.7, after consultation with the contractor, fix a later date for the contractor to comply with clause 35.7 *or* • if the contractor has stated that non-compliance is due to other things stated in the notice, write to the contractor either: – that the things do not justify non-compliance and the contractor must comply *or:* – that the things do justify non-compliance and issue further instructions to enable the contractor to comply or cancel the nomination and omit the work or nominate another sub-contractor. Send copy of instruction to the sub-contractor.
35.13.1	**Duty** When issuing each interim certificate, direct the contractor regarding the amount included for each nominated sub-contractor. **Duty** Compute the amounts in accordance with NSC/C. Inform each nominated sub-contractor forthwith of the amount.
35.13.5.1	**Duty** Issue a certificate that the contractor has failed to provide reasonable proof of payment of a nominated sub-contractor with a copy to the nominated sub-contractor.
35.14.2	**Duty** Operate the provisions of NSC/C regarding sub-contract extensions of time.
35.15	**Duty** If the nominated sub-contractor has failed to complete the sub-contract works by the date in the sub-contract, if the contractor notifies the architect, if the architect is satisfied that the extension of time procedures have been properly done, certify in writing to the contractor with a copy to the nominated sub-contractor that the nominated sub-contractor has so failed.

Table 2.1 *Contd*

Clause	Power/Duty
35.16	**Duty** Certify practical completion of the nominated sub-contract works when it has been achieved and NSC/C clause 5E.5 has been complied with sufficiently. Send a copy to the nominated sub-contractor. Send a copy of a written statement under clause 18.1 if appropriate.
35.17	**Power** Issue an interim certificate including the final amount due to the nominated sub-contractor, if NSC/W clause 5 applies, at any time after certification of practical completion of the nominated sub-contract works. **Duty** Issue an interim certificate including the final amount due to the nominated sub-contractor, if NSC/W clause 5 applies, on the expiry of 12 months from certification of practical completion of the nominated sub-contract works.
35.18.1.1	**Duty** Issue an instruction nominating a substituted sub-contractor, if the contractor has fully paid the nominated sub-contractor who fails to remedy defects.
35.24.6	**Duty** Issue an instruction to the contractor to give the nominated sub-contractor a default notice prior to determination.
35.24.6.3	**Duty** Make further nomination if the contractor informs the architect that the nominated sub-contractor's employment has been determined.
35.24.7.2	**Power** Without his consent to determination in the case of insolvency if the contractor and architect are reasonably satisfied that the administrator or the sub-contractor is willing and able to carry out the sub-contract. **Duty** If the architect is not so satisfied, consent to the determination unless the contractor and employer otherwise agree.
35.24.7.3	**Duty** Make such further nomination as necessary.
35.24.7.4	**Duty** Make such further nomination as necessary.
35.24.8.1	**Duty** Make such further nomination as necessary.
35.24.8.2	**Duty** Make such further nomination as necessary.
35.26.2	**Duty** Issue an interim certificate including the value of work and materials to the extent not previously certified where the nominated sub-contractor's employment is determined.
36.2	**Duty** Issue instructions nominating a supplier.

Clause	Power/Duty
	Table 2.1 *Contd*
36.3	**Duty** Add expense to the contract sum if the architect's opinion is that the nominated supplier has incurred expense not reimbursable under clause 36.3.1.
36.5.2	**Power** Approve in writing, restrictions, limitations or exclusions in any contract of sale.
42.5	**Power** Within 14 days of receiving the contractor's statement, send written notice to the contractor requiring him to amend his statement if it is insufficient to explain the proposals for performance specified work.
42.6	**Duty** Immediately give notice to the contractor if the architect finds a deficiency which would adversely affect the performance required.
42.11	**Power** Issue instructions requiring a variation to the performance specified work.
42.13	**Power** Require the contractor to provide an analysis of the performance specified work part of the contract sum.
42.14	**Duty** Issue instructions for the integration of performance specified work.

2.2   The architect as agent for the employer

When an architect is certifying under the contract, he is exercising his professional judgment independently of his employer and not as agent for the employer: *Sutcliffe* v. *Thackrah* (1974). However, in some instances, the architect may be regarded solely as the agent of the employer as was the situation in the case of the city architect in *Rees and Kirby Ltd* v. *Swansea City Council* (1983). He was the named architect under a JCT 63 contract between Rees and Swansea for the erection of a housing estate. The architect failed to certify on a claim for loss and/or expense. The judge said:

'The more interesting point is whether the architect was agent for the defendants, making them vicariously liable...
As I understand the law an architect is usually and for the most part a specialist exercising his special skills independently of his employer. If he is in breach of his professional duties, he may be sued personally.

There may, however, be instances where the exercise of his professional duties is sufficiently linked to the conduct and attitude of the employer that he becomes the agent of the employers, so as to make them liable for his default.

In the instant case, the employers, through the behaviour of the Council and the advice and intervention of the Town Clerk, were to all intents and purposes dictating and controlling the architect's exercise of what should have been his purely professional duty.

In my judgment, this was the clearest possible instance of responsibility for the breach attaching to the employers.'

It is not unknown for a local authority to refuse to allow its employed architects to issue final certificates until the council's audit department have satisfied themselves about every particular in a final account prepared by the quantity surveyor. Similar pressure is sometimes brought on independent consultants to await the results of an audit before certifying. Such councils may make themselves vicariously liable, at least in the case of employed architects. Interfering with the proper exercise of the duty of a truly independent consultant raises still further issues. In such cases, the independent consultant must ask for clear instructions. If the instructions are to await the results of the audit, it probably amounts to a repudiation on the part of the employer which the consultant has little choice but to accept. The alternative probably amounts to him conniving in breach of contract.

The employer will be estopped from denying that the architect has his actual authority to do those things which are specified for him to do in JCT 98. To take but one instance: variations. An architect has no implied authority to vary the work contracted, because the Standard Form of Agreement for the Appointment of an Architect (SFA/99) clause 2.7 expressly states that the architect must not make any material alteration or addition or omission in regard to services or approved design without, not only the knowledge but also, the consent of the client, except in the case of an emergency. But the contractor does not have to worry whether or not the architect actually has the consent of his client for any instructions issued within the terms of JCT 98; the employer will be estopped so far as the contractor is concerned from denying that such variations were made with his consent.

Failure to do what the contract requires of an architect may serve as evidence of negligence in an action by the contractor in tort against the architect. It appears also that the architect owes a duty in tort to see that a contractor or sub-contractor does not suffer loss by his negligence.

2.3 Architect's duties to the employer

In spite of its length, the Standard Form of Agreement for the Appointment of an Architect 1999 (SFA/99) is incomplete in the sense that there are other duties which the law will imply on the architect in the performance of his professional obligations.

These duties will apply to anyone who carries out the work normally ascribed to an architect, whether or not that person is in fact registered with the Architects Registration Board (ARB) established as the successor body to the Architects Registration Council of the United Kingdom (ARCUK) by the Housing Grants, Construction and Regeneration Act 1996 and governed by the Architects Act 1997. Although no one in the United Kingdom may refer to himself as an 'architect' for business purposes unless registered with ARB, anyone can carry on the business of architecture without registration provided he does not refer to himself as an 'architect'. Many people do act in this way and find themselves fulfilling the architect's role under the contract albeit under the title of 'Contract Administrator'.

However, the fact that a person is not on the register does not impose on him lesser duties than those which are imposed on architects. In *Oxborrow* v. *Godfrey Davis (Artscape) Ltd and Others* (1980), a firm of estate agents prepared plans for a bungalow which were drawn up by one of their employees, calling himself an 'architectural draftsman'. When strip footings which he had designed proved to be inadequate for a clay site with trees in proximity to the bungalow, the estate agents were held to be just as liable as if an architect had designed them. In the event, the liability was distributed equally between the estate agents and the builder, the local authority having 20% liability. In *E.H. Cardy & Son Ltd* v. *A.E.V. Taylor and Paul Roberts & Associates* (1994) the court applied the same stringent rules to someone crassly described as an 'unqualified architect' as they would have done to a registered architect.

It is outside the scope of this book to examine in detail all the architect's duties, but they include the duty to advise the employer about the law so far as it affects the Works, the appropriate procurement route and contract form and the proper completion of that form. Although he is not required to have the expert knowledge of a lawyer, he is expected to have a command of the law applicable to planning, building regulations and applicable statutes: *Townsend Ltd* v. *Cinema News, David A. Wilkie & Partners (Third Parties)* (1959); and the rights of adjoining occupiers and owners, including rights to light: *Armitage* v. *Palmer* (1960). As with other professional men, he will be expected to keep abreast of developments in the law, including recent legislation and court decisions and he will be liable in damages for negligence to his client if he does not do so: *B.L. Holdings Ltd* v. *Wood* (1979). In particular, he will be expected to have a sufficient knowledge of the main forms of contract to advise his client about the most suitable form for a particular project, and a detailed knowledge of the JCT 98 contract if that is used. Sadly, many architects and quantity surveyors become so used to a particular form of contract that they will use no other even when their favourite contract is clearly unsuitable. If the employer suffers a loss as a direct result, the architect or quantity surveyor will be liable in negligence. In *Terry Pincott* v. *Fur & Textile Care Ltd* (1986), it was held that it was part of the duty of a competent architect to:

> '…advise his client and not mislead the contractor, as to the nature of any agreement that he recommends should be entered into between the client and the contractor.'

The architect was held to be negligent for failing to understand correctly MW 80 and had incorrectly advised both his client and the contractor of the effect of it. He was also held liable for failing to obtain planning permission for the work.

However, the architect's duties do not seem to include advising his clients on the commercial feasibility or marketability of any project: *Eli Abt* v. *Brian Fraiman* (1975). But an architect may be liable in damages for negligently advising his client of the letting area of a building: *Gable House Estates Ltd* v. *Halpern Partnership & Another* (1995).

An architect must bring to his task the care and skill usual among averagely competent architects practising their profession. The standard of care of a professional person is generally accepted to have been laid down by Mr Justice McNair in *Bolam* v. *Friern Hospital Management Committee* (1957), a medical case.

An architect's obligation to use due care and skill without doubt includes some element of due diligence.

An architect who grants unwarranted extension of time under JCT 98 to a contractor will be liable in damages to his client even if the client has obtained damages in arbitration against the contractor in respect of the extensions of time and for defective work: *Wessex Regional Health Authority* v. *HLM Design Ltd (No 2)* (1994); see also *Royal Brompton Hospital National Health Service Trust* v. *Frederick Alexander Hammond & Others* (1999) where the judge memorably said that the duty of a professional man, generally stated, is not to be right but to be careful.

Any architect will be liable to his client if he causes loss by his negligence: *Sutcliffe* v. *Thackrah* (1974) and the client can sue the contractor and the architect in any order he wishes (subject to the existence of an arbitration clause).

The liability of architects, of course, extends to their employees and one has been held negligent for employing an assistant who was not properly qualified: *Alexander Corfield* v. *David Grant* (1992).

An architect may be liable in damages to his client if he fails to give the appropriate default notice prior to termination of a contractor's employment because he has failed to proceed regularly and diligently with the work: *West Faulkner Associates* v. *London Borough of Newham* (1995).

The architect has a duty to warn a client before accepting a tender of the contractor if the contractor is not financially sound. The *Architect's Job Book* published by RIBA Publications requires the architect to discreetly check the financial status of firms. Checks to be made should include enquiries of builders' merchants, banks and trade credit references, a company search, and enquiries of other building professionals. Failure to make enquiries is a breach of the architect's obligation to his client: *Partridge* v. *Morris* (1995).

If the architect condones the contractor's deviation from the agreed plans, however minor, this is negligent: *Parochial Church Council of Holy Trinity Church Much Wenlock* v. *Leonard Baart* (1993).

Lord Justice Bingham in *Watts* v. *Morrow* (1991) said: 'A contract breaker is not in general liable for any distress, frustration, anxiety, vexation, tension or aggravation which his breach of contract may cause to the innocent party ...'. But the rule is not absolute. Where the very object of the contract is to provide pleasure, relaxation, peace of mind or freedom from molestation, damages will be awarded if the purpose of the contract is not provided or if the contrary result is procured instead. But in *Maurice L. Knott and Another* v. *Terence P. Bolton and Another* (1995), the defendants designed a house for the plaintiffs who had provided a sketch plan and emphasised that they required a grand hall and staircase. During the course of construction, it was discovered that the staircase and gallery structure could not be fitted into the hall. The staircase itself was only 85 cm wide. The plaintiffs were not aware of the problems until it was too late to do anything about them. The plaintiffs brought an action for damages under a number of headings, among them was 'distress'. This head was rejected by the judge who held that unless the purpose of the contract was to confer pleasure on the plaintiffs, this claim must fail. The plaintiff's appeal was dismissed.

Recently, it has been held that, subject to whatever his terms of agreement may say, an architect is not normally under a duty to review his design after practical completion: *New Islington & Hackney Housing Association Ltd* v. *Pollard Thomas & Edwards Ltd* (2001).

An extensive list of an architect's duties from the employer's point of view was produced by Alfred Hudson and is now to be found in *Keating on Building Contracts* (2000) (7th edition) at page 393. It is still applicable in general terms and it may be very useful where there is no formal contract and the question of implied terms arises. But of course the relationship between architect and client is principally determined by the contract between them. Indeed, the Codes of Professional Conduct of both the RIBA and ARB require an architect to confirm the terms of appointment in writing to his client.

2.4   Power to postpone work

The architect is empowered under clause 23.2 to issue instructions in regard to the postponement of any work to be executed under the provisions of this contract.

The words, on the face of it, may allow the architect to veto the order in which the contractor proposes to do any work, but it is one to be used sparingly since the contractor may be entitled to an extension of time under clause 25.4.5.1 and loss and expense under clause 26.2.5, if the 'regular progress of the Works or any part thereof' has been materially affected; and it may even lead to determination of the contract by the contractor under 28.2.2.2.

Instructions to postpone under this clause may be given, it has been said, in effect by implication. An architect's nomination of a sub-contractor on the basis of that sub-contractor's 'quotation ... and tendering conditions', which terms involved delay in the contractor's original planned programme of work, was held to amount to a postponement under JCT 63 clause 12(2) (a clause similar in effect to clause 23.2 of JCT 98): *Harrison* v. *Leeds* (1980). Lord Justice Megaw said:

'While the ... order does not expressly use the word "postpone" or expressly give instruction to postpone, the order to make a contract with sub-contractors containing a condition which necessarily involves postponement is an instruction within the contemplation of the condition [amounting to an order to postpone].'

The arbitrator had originally found that the order did not constitute a postponement order and that view has much to commend it. The contractor's intended, but not expressed, order in which he would carry out the Works may have been the basis of his tender, but that was in no sense part of the contractual obligation he undertook; still less was his programme, even if the architect had seen it, a contract document. The arbitrator found as a fact: 'The claimants prepared an outline programme ... but this programme did not form part of their tender'.

If the decision of the Court of Appeal in this case is correct, it seems that any instructions which result in alteration to some optimistic programme, prepared in secret by a contractor before tendering, may amount to implied instructions by the architect to postpone. That seems improbable since JCT 98 (as did JCT 63) entitles the contractor to loss and/or expense and, in some circumstances, even power to determine the contractor's employment for postponement of the Works. To place this interpretation on JCT 98 clause 23.2 is to ignore the words: 'under the provisions of this contract'. It is thought that postponement must mean the stopping of work which is actually in progress or about to start rather than referring to some notional programme which may not be achieved. It should also require a formal order and not be a mere incident or consequence of some other instruction as seems to have been the position in *Holland Hannen and Cubitts* v. *Welsh Health Technical Services Organisation* (1981). In that case, an architect issued an instruction that the contractor was to ensure that all windows were weather-tight before any internal finishes were applied. The judge considered the instruction to have been a postponement instruction in effect although it seems the architect was simply insisting that the windows be weather-tight. The treatment of defective work is something about which the architect must take particular care.

2.5   Power to have defective work removed

If the contractor fails to carry out the work in accordance with the contract documents, the architect will want to instruct him to rectify it. Strangely, the contract does not invest the architect with such power. Instead, JCT 98 clause 8.4 contains an elaborate series of sub-clauses regarding defective work. Clause 8.4.1 is the main clause which provides that the architect may issue instructions in regard to the removal from site of any work, materials or goods which are not in accordance with the contract. Although an instruction dealing with defective painting in terms of 'Remove from site paintwork not in accordance with the contract' conjures up visions of operatives painstakingly shaving paint from woodwork and wheeling it off site in barrow loads, it is the only form of instruction regarding defective work which the contract recognises.

The point was considered by the court in *Holland Hannen and Cubitts* v. *Welsh Health Technical Services Organisation* (1981). The architect issued notices which purported to be under clause 6(4) of JCT 63 (equivalent in wording to 8.4.1 of JCT 98) condemning the window assemblies. Judge John Newey said:

> 'The first of these documents was in letter form and did not mention clause 6(4). The second and third were more formal and referred to the sub-clause. None of them, however, instructed Crittalls to remove the window assemblies from the site. The effect of doing so would have been to expose the interior of the hospital to wind and rain ... In my opinion, an architect's power under clause 6(4) is simply to instruct the removal of work or materials from the site on the ground that they are not in accordance with the contract. A notice which does not require removal of anything at all is not a valid notice under clause 6(4).'

This is a startling, but strictly literal, interpretation of the wording. But if a notice is given which is expressed to be under clause 8.4.1, condemning the windows, it ought to imply, let alone the application of simple common sense, that the contractor is to remove the condemned items if that is the only way to rectify the defects. The judge went on to say:

> 'I think that the three purported notices in this case were all invalid and of no effect under the contract. If I am wrong and the notices are to be construed as having required removal of the window assemblies, then they were instructions which were not obeyed. Since [the employers] did not invoke their sanction under clause 2(1) of bringing other contractors to remove the windows, their effect has long since spent.'

It would appear, according to this court, that if an architect issues a notice under clause 8.4.1 which is a valid notice, but which is ignored, and no further action is taken beyond allowing the contractor to attempt remedial work, this acquiescence will in time deprive the notice of all effect. That is a surprising conclusion to say the least.

What is clear is that an architect purporting to issue an instruction under clause 8.4.1 must take care to phrase the instruction strictly in accordance with the wording of the clause and in particular requiring 'removal from site' of the defective work rather than rectification or making good.

Clause 8.4 has three further sub-clauses which add an extra dimension to the architect's powers to deal with defects:

2.5.1   Clause 8.4.2

If the employer agrees, the architect may 'consult' with the contractor who must 'consult' with any relevant nominated subcontractor, before allowing any defective materials, goods or workmanship to remain on site. This must be confirmed in writing. The obligation to consult has little practical significance. The literal meaning is to 'seek advice or information', but there is no requirement that any of the participants are to do anything with such advice or information. In practical terms, it means no more than that, if the architect decides that the best option may be to leave the defect in the Works, he must get the employer's consent and discuss the matter with the contractor, hopefully (but not essentially) getting the contractor's agreement. The clause expressly states that the architect's written confirmation is not to be taken to be a variation. In order to achieve the full protection from this provision, the architect must ensure that the confirmation makes specific reference to this clause. The sting in the tale for the contractor is that an 'appropriate deduction' must be made from the contract sum. This phrase is used in clause 17 **[12.4]**, but here it is clear that an appropriate deduction could well be greater than the rate in the bill of quantities. The quantity surveyor must take into account all the circumstances before deciding what is the true value of the defective work or materials.

2.5.2   Clause 8.4.3

The architect is entitled to issue variation instructions 'as are reasonably necessary' resulting from an instruction to remove defective work or indeed allowing it to remain in place. In the first instance, it will be for the architect to decide whether the instruction is reasonably necessary. If the contractor disagrees, he can refer the dispute to adjudication or whichever of the arbitration or litigation options have been chosen. Again the fairly meaningless process of consultation is set out as a prerequisite. Very importantly, no addition is to be made to the contract sum nor any extension of time given 'to the extent' that the instruction is reasonably necessary. Therefore, if the instruction is partially, but not entirely necessary, the valuation, loss and/or expense and extension of time provisions of the contract have to be applied to the part of the instruction that is not reasonably necessary. Despite this necessary limitation on the architect's power, this is a most useful clause in appropriate circumstances.

2.5.3   Clause 8.4.4

After the power to order removal of defective work from site, clause 8.4.4 is potentially the most useful power given to the architect to deal with defective work. It enables him to instruct the contractor to open up work or to test work or materials following an instruction to remove defective work. The architect can order the opening up or testing to reasonably satisfy himself that there are no other similar defects. Again, no addition is to be made to the contract sum nor any extension of time given 'to the extent' that the instruction is reasonable in all the circumstances. That is the case even if the opening up or testing reveals that the work inspected is fully in accordance with the contract. The logic behind this is clearly that the architect's actions are triggered by the discovery of a defect. Therefore, if the instruction is partially, but not entirely necessary, the valuation, loss and/or expense and extension of time provisions of the contract have to be applied to the part of the instruction that is not reasonably necessary. An important proviso is that the architect may not exercise his power until he has 'had due regard' to the Code of Practice which is included in the contract after clause 42. The requirement to have regard to something can be a difficult concept. It obviously means that the architect must pay attention to the Code, but not that he must slavishly follow it: *R* v. *Greater Bir-*

mingham Appeal Tribunal, ex parte Simper (1973). The Code states that the architect and the contractor should try to agree the amount of opening up and testing and lists 15 criteria. Perhaps the most telling is the fifteenth item which is: 'any other relevant matters'.

Clause 8.3 should be noted. It differs markedly from clause 8.4.4, because although it also gives the architect power to order the opening up or testing of work and materials, it carries a heavy penalty for the employer if the architect is mistaken. It does not depend on a discovered defect to trigger the instruction. It is enough that the architect gives the instruction. Usually, of course, the architect will have a good reason for believing it to be necessary to give an instruction under this clause – a suspicion that certain damp-proofing materials have been omitted, for example. If the investigation reveals that the work or materials are not in accordance with the contract, the contractor is obliged to rectify the situation at his own cost and he gets neither extra time nor money to cover the investigation or the correction. If, however, the work and materials are found to be in accordance with the contract, the contractor will be entitled to be paid the cost of investigation and reinstatement and be given an extension of time if warranted. If he also suffers loss and/or expense, he will be entitled to make application in accordance with clause 26 **[10.11.2]**.

The remedies under clause 8.4 are expressed to be without prejudice to the generality of the architect's powers. Although impressive, it is in reality an empty phrase, because the contract nowhere invests the architect with any general powers. Indeed, even clause 4 does not give the architect any general power to issue instructions, but only such instructions in regard to any matter 'in respect of which the Architect is expressly empowered by the Conditions to issue instructions...'.

Clause 8.5 deals with the architect's power to issue instructions to deal with the contractor's failure to carry out the work in a proper and workmanlike manner. Again, no addition is to be made to the contract sum nor any extension of time given 'to the extent' that the instruction is reasonably necessary.

Clause 8.6 empowers the architect to issue instructions to exclude from site any person employed. The instruction must not be issued unreasonably or vexatiously **[9.3]**.

2.6   No power to direct contractor

'It is the function and right of the contractor to carry out his building operations as he thinks': *Clayton v. Woodman & Sons Ltd* (1962). The architect has no power to tell a contractor how to do the work or in what sequence: *Greater London Council* v. *Cleveland Bridge* (1984).

In the absence of specific contractual obligations to the contrary, an architect is not under a duty to his employer, still less does he have any right to direct a contractor how or when he is to do the work. He is not required or allowed to supervise the contractor's method of working, unless the specification or other parts of the contract require that the contractor shall adopt certain methods or timing: *Clayton* v. *Woodman & Son (Builders) Ltd* (1962); *AMF International Ltd* v. *Magnet Bowling Ltd* (1968); *Sutcliffe* v. *Chippendale and Edmondson* (1971).

Clause 13.2 empowers the architect to issue instructions requiring a variation, and clause 13.1 defines what a variation is **[6.2]**. One of the possible variations is described in clause 13.1.2 as the imposition, addition or omission, or the alteration of any obligations or restrictions already imposed in the bills of quantities, by the employer. The type of obligations are listed as access or use of specific parts of the site, limitation of working space or hours and the carrying out or the completion of the work in any particular order. It may be argued that this amounts to directing the contractor, albeit in a wholly negative way.

The contractor is entitled to object to an instruction issued under this clause. The entitlement is conferred by clause 4.1.1. He need not comply with a variation instruction 'within the meaning of clause 4.1.1' to the extent that he makes reasonable written objection. Is it a reasonable objection that he would not make as much money as he otherwise would have made? That appears to be an eminently reasonable objection.

If the architect sees that the contractor has taken a wrong course which will result in injury to property or individuals, he has an obligation to warn them: *Oldschool* v. *Gleeson (Construction) Ltd* (1976); but he has no power to do more than that.

2.7   *Does the architect owe a duty of care to the contractor?*

This is an extremely complex question. In *Michael Sallis & Co Ltd* v. *Calil* (1987), the contractor sued Mr & Mrs Calil and the architects, W F Newman & Associates. It was claimed that the architects owed a duty of care to the contractor. The claim fell into two categories:

• Failure to provide the contractors with accurate and workable drawings

- Failure to grant an adequate extension of time and under certification of work done.

The court held that the architect had no duty of care to the contractors in respect of surveys, specifications or ordering of variations, but he did owe a duty of care in certification. It was held to be self-evident that the architect owed a duty to the contractor not to negligently under-certify.

> 'If the architect unfairly promotes the building employer's interest by low certification or merely fails properly to exercise reasonable care and skill in his certification it is reasonable that the contractor should not only have the right as against the owner to have the certificate revised in arbitration but also should have the right to recover damages against the unfair architect.'

In arriving at that conclusion, the court was following the rules laid down by many courts. In *Campbell* v. *Edwards* (1976), the Court of Appeal said that the law had been transformed since the decisions of the House of Lords in *Sutcliffe* v. *Thackrah* (1974) and *Arenson* v. *Arenson* (1975); because contractors now had a cause of action in negligence against certifiers and valuers. Before these cases, certifiers had been protected because the Court of Appeal in *Chambers* v. *Goldthorpe* (1901) had held that certifiers were quasi-arbitrators. The House of Lords overruled that in 1974; until the *Pacific Associates* case (1988), at no time in the history of English law had it been doubted that architects owed a duty to contractors in certifying. After all, there was no need even to invent the doctrine of quasi-arbitrators if there was no liability for negligence. In the *Arenson* case in reference to the possibility of the architect negligently under-certifying, it was said:

> 'In a trade where cash flow is perceived as important, this might have caused the contractor serious damage for which the architect could have been successfully sued.'

Doubt was cast upon the centuries old law about the architect owing a duty to the contractor in certification by *Pacific Associates* v. *Baxter* (1988). Halcrow International Partnership were the engineers for work in Dubai for which Pacific Associates were in substance the contractors under a FIDIC contract. In the course of the work, the contractor claimed to have encountered unexpectedly hard materials for which they were entitled to extra payment of

£31 million. Halcrow refused to certify and in due course, Pacific Associates sued them for £47 million, being the £31 million plus interest and another item less credit. It was claimed that Halcrow acted negligently in breach of their duty to act fairly and impartially in administering the contract. At first instance, the court struck out the claim, holding that Pacific Associates had no cause of action. Two points impressed the court:

- There was provision for arbitration between employer and contractor
- There was a special exclusion of liability clause in the contract (clause 86) to which, of course, the engineers were not a party, whereby the employers were not to hold the engineers personally liable for acts or obligations under the contract, or answerable for any default or omission on the part of the employer.

The existence of an exclusion of liability clause is not relevant to the question of whether a duty of care exists, except to the extent that the existence of such a clause suggests acceptance by the engineer that there is a duty of care which, without such a clause, would give rise to such liability. It is, at best, doubtful whether such a clause would be deemed reasonable under the provisions of the Unfair Contract Terms Act 1977: *Smith* v. *Eric S. Bush* (1989). Even more surprisingly it was held that the inclusion of an arbitration clause in the contract, General Condition 67, excluded any liability by the engineer to the contractor. The rationality of that is not immediately apparent. If the employer and the contractor choose to settle any disputes by arbitration rather than litigation, why should that exonerate a third party from the obvious duty to both? However, it seems that it was these two points which were decisive.

The Court of Appeal upheld the decision, which can be criticised on three major bases:

(1)   The Court of Appeal is bound by its own previous decisions. In *Lubenham Fidelities* v. *South Pembrokeshire District Council* (1986) it expressly affirmed the principle that the architect owes a duty to the contractor in certifying. The architects in that case were not held liable, because the chain of causation was broken and the contractor's damage was held to be caused by their own breach in wrongfully withdrawing from site. But the court said:

'We have reached this conclusion with some reluctance, because the negligence of Wigley Fox [the architects] was undoubtedly the source from which this unfortunate sequence of events began to flow, but their negligence was overtaken and in our view overwhelmed by the serious breach of contract by Lubbenham.'

It expressly approved the first instance judgment where it was said:

'Since Wigley Fox were the architects appointed under the contracts, *they owed a duty to Lubbenham as well as to the Council to exercise reasonable care in issuing certificates and in administering the contracts correctly.* By issuing defective certificates and in advising the Council as they did, Wigley Fox acted in breach of their duty to Lubenham.' (author's emphasis)

(2) It apparently overruled all the previous cases, including those of the House of Lords by which it was bound, going back for more than a century, together with well-established law that had been followed in all common law jurisdictions such as Hong Kong and Australia.

(3) It ignored the fundamental principle that (at that time) parties could not be bound by a term in a contract to which they were not a party and had not consented.

Recent cases such as *Henderson* v. *Merritt Syndicates Ltd* (1994) and *Conway* v. *Crowe Kelsey* (1994) suggest that the reliance principle established in *Hedley Byrne & Co Ltd* v. *Heller and Partners Ltd* (1963) is capable of extension to accommodate actions as well as advice given by the architect. In *J. Jarvis & Sons Ltd* v. *Castle Wharf Developments & Others* (2001) it has been held by the Court of Appeal that a professional who induces a contractor to tender in reliance on the professional's negligent misstatements could become liable to the contractor if it could be demonstrated that the contractor relied on the misstatement. There is a very perceptive article by John Cartwright (Liability in Negligence: New Directions or Old) in *Construction Law Journal* (1997) vol. 13, page 157.

2.8 Correcting discrepancies

Clause 2.2.2.1 provides that the bills of quantities are to have been prepared in accordance with the standard method of measurement (SMM7). Clause 2.2.2.2 deals with discrepancies between the bills of quantities and SMM7 or in descriptions, quantities or omitted items and provides that they must be corrected and such corrections treated as variations. Although the architect is not specifically mentioned, it is clearly he who must do the corrections.

Clause 2.3 deals with other kinds of discrepancies between the contract drawings, the bills and architect's instructions. Clearly, none of these provisions has any reference to a master programme prepared by the contractor and delivered to the architect in accordance with clause 5.3.1.2 and it seems that any discrepancies in the master programme may lie uncorrected.

Clause 2.3 provides that on notification by the contractor, the architect must issue instructions about the discrepancy. It does not expressly state that he must resolve the discrepancy or that he must resolve the matter by choosing one of two or more discrepant items, but he must resolve the discrepancy in some way. Within that broad parameter he appears to have some discretion.

There is no provision for the correction of contractors' errors in pricing or in multiplication or addition. A contractor who mistakenly submits rates which are too low to enable him to make a profit has no escape. On the other hand, if he deliberately prices a small quantity at a high rate in the hope of an increase in the quantity and, therefore, a sizeable profit, that is acceptable pricing strategy and the employer has no redress: *Convent Hospital* v. *Eberlin & Partners* (1988).

If a contractor submits a tender containing a mistake against his interests and the employer or his architect or quantity surveyor sees it, deliberately keeps quiet and takes advantage of it, a court may correct the error in the exercise of its equitable jurisdiction and on the basis of estoppel: *Riverlate Properties Ltd* v. *Paul* (1974); *McMaster University* v. *Wilchar Construction Ltd* (1971).

2.9   Architect's instructions

Clause 4.1.1 makes clear that the contractor must comply with all instructions issued by the architect. The provision is not all-embracing, however, because apart from variation instructions issued under clause 13.1.2 **[6.2]**, the contractor need only comply if the instruction is expressly empowered by the contract. It might then be expected that the clause would list all such instructions as does the government contract (GC/Works/1(1998)) or the Association of Consultant Architects contract (ACA 3). Not so; the architect and the contractor are left to read through more than one hundred pages of the document in order to locate the information. The instructions referred to are listed in Table 2.2.

Table 2.2
Architect's instructions empowered by the JCT 98 contract

Clause	Instruction
2.3	After being notified in writing of a discrepancy in documents.
2.4.1	After being notified in writing of a discrepancy or divergence between the contractor's statement in connection with performance specified work and an architect's instruction.
6.1.3	Relating to any divergence between statutory requirements and the contract documents or any instruction notified by the contractor or discovered by the architect.
6.1.6	Relating to any divergence between statutory requirements and the contractor's statement in connection with performance specified work if the contractor has informed the architect in writing of proposals for rectifying the divergence.
7	Not to amend errors in setting out if the employer consents.
8.3	For the opening up or testing of work or materials.
8.4.1	Regarding the removal from site of work not in accordance with the contract.
8.4.3	Requiring reasonably necessary variations resulting from defective work.
8.4.4	To open up or test as is reasonable in all the circumstances after having had regard to the Code of Practice to the contract to establish the likelihood of similar non-compliance.
8.5	Requiring reasonably necessary variations or otherwise resulting from the contractor's failure to carry out the work in a workmanlike manner.
8.6	Requiring the exclusion from site of any person there employed.
13.2.1	Requiring a variation.
13.2.3	That a variation is to be carried out and valued under clause 13.4.1.
13.3	About the expenditure of provisional sums in the bills of quantities and in nominated sub-contracts.
13A.4	That a variation is to be carried out and valued under the default procedure or that the variation is not to be carried out, if the employer does not accept the clause 13A quotation.

Table 2.2 *Contd*

Clause	Instruction
17.2	Specifying a schedule of defects not later than 14 days after the end of the defects liability period. Not to make good defects if the employer consents.
17.3	The making good of defects whenever he considers it necessary during the defects liability period. Not to make good the defects if the employer consents.
21.2.1	To take out insurance in joint names in respect of liability for collapse, subsidence, heave, vibration, weakening or removal of support or lowering of ground water.
22D.1	To obtain a quotation. Whether or not the employer wishes the contractor to accept the quotation.
22FC.3.1.2	To enable compliance to the extent that the remedial measures require a variation.
23.2	Postponing work.
27.6.3	Requiring the contractor remove from the Works temporary buildings, plant, tools, etc.
30.6.1.1	To provide the quantity surveyor with all documents necessary for the adjustment of the contract sum.
34.2	Regarding antiquities reported by the contractor.
35.5.2	To remove the contractor's objection to a proposed nominated sub-contractor. To omit the work. Nominating another sub-contractor.
35.6	Nominating a sub-contractor.
35.18.1.1	Nominating a substituted sub-contractor.
35.24.6	To give the nominated sub-contractor a default notice prior to determination.
36.2	Nominating a supplier.
42.11	Requiring a variation to the performance specified work.
42.13	Requiring the contractor to provide an analysis of the performance specified work part of the contract sum.
42.14	For the integration of performance specified work.

Clause 4.3.1 requires that all instructions issued by the architect must be in writing. It then goes on to give elaborate provisions to cover the situation where the architect only gives oral instructions. Clause 4.3.2 allows the contractor to confirm them within seven days and if within a further seven days from receipt of the notice the architect has not dissented from them, such instructions are effective not from the date of the original oral instruction, surprisingly, but from the date which is seven days from the date on which the architect received the contractor's notice.

The contractor is not obliged to confirm in writing if the architect issues a written confirmation within seven days (clause 4.3.2.1). It does not take a genius to calculate that if the contractor waits to see if the architect will confirm in writing within the seven days allotted, the contractor will miss his own opportunity to confirm. A sensible contractor will not wait, but will confirm all oral instructions from architects promptly. Most contractors seem to have a standard form which they call, erroneously, a CVI (confirmation of verbal instruction). 'Verbal' simply means 'relating to words'; it does not mean 'word of mouth' as many in the construction industry appear to believe. An 'oral' instruction is one given by mouth i.e. spoken. Where the architect confirms, the instruction is effective from the date of confirmation.

The contract goes on to provide in clause 4.3.2.2 that if neither architect nor contractor confirms in writing, but the contractor nevertheless complies with the instruction, the architect may (but is not obliged to) confirm them in writing at any time up to the issue of the final certificate. With masterly inconsistency, the instruction is then stated to be effective from the date of the original oral instruction. Therefore, the position would be that the later the architect decides to confirm, the earlier the effective date of the instruction. Clause 13.2.4 appears to enlarge this power by giving the architect right to 'sanction in writing' a variation carried out by the contractor even if not instructed by the architect. So if the contractor carries out work which is not in accordance with the contract, the architect may effectively bring the work within the contract by using the sanction and then it is to be valued in the appropriate way **[6.4]**.

What is the significance of the date on which the instruction becomes effective? If both architect and contractor fail to confirm an oral instruction and, therefore, the contractor, quite properly, does not comply but the architect confirms the instruction much later, the instruction is deemed to have taken effect on the date it was orally issued. Theoretically, this leaves the contractor open to accusations of breach of his obligation to comply with an instruction. It would, however, be a very brave, not to say misguided, architect who was prepared to pursue such a distinctly unmeritorious argument.

If the contractor has neither an instruction in writing before he does the work nor a confirmation by himself or the architect, he has no right to payment under the terms of the contract. In spite of this, contractors commonly carry out architect's oral instructions. They say, often with good reason, that if they waited for confirmation of the instruction before complying, the project would be seriously delayed. There never was any excuse for oral instructions. There is always some means by which the architect can record the instruction. Even if the instruction is given over the telephone, the architect can easily fax written confirmation within minutes. Similarly if the e-mail option is chosen **[1.13]**. In certain circumstances, the employer may be held to have waived the requirement for instructions to be in writing if instructions are routinely given orally. In such circumstances, he may not be able to refuse payment: *Bowmer & Kirkland* v. *Wilson Bowden Properties Ltd* (1996); *Redheugh Construction Ltd* v. *Coyne Contracting Ltd and British Columbia Building Corporation* (1997).

If the contractor fails to comply with an architect's instruction, the architect has quite draconian powers. If the instruction is in relation to the removal of defective work from site, the architect may, subject to certain provisos, serve a notice of default leading to determination of the contractor's employment if the default is not rectified **[9.4]**. Otherwise, under clause 4.2, the architect may issue a notice requiring the contractor to comply with his instruction within seven days. If the contractor does not comply, the employer may engage another contractor to carry out the work necessary to comply with the instruction. The power is very broadly drafted and proceeds to state that 'all costs incurred in connection' with employing the other contractor may be deducted by the employer from money due or be recoverable as a debt. Many contractors and architects do not appreciate the wide nature of such costs. They clearly encompass the costs of obtaining competitive prices from other contractors together with any consultants' fees. The obvious intention is that the employer should not suffer any financial loss due to the contractor's non-compliance.

If the employer chooses to deduct the costs from a certified amount, by far the easiest option, he must issue the appropriate withholding notices **[5.21]**. If he wishes to recover the costs as a debt, perhaps because there is not enough money owing to the contractor from which to make a deduction, he may sue for it through the courts or arbitrate depending on the dispute resolution option chosen **[11.2]**. The architect has no power to simply adjust the contract sum to take account of the costs.

2.10   *Instruction issued after the date of practical completion*

A question which arises from time to time is whether the architect has power to issue an instruction after the date of practical completion (whether the architect can issue an instruction requiring a variation after the contract date for completion is discussed in section 7.18). Clearly the architect is entitled to issue certain instructions; for example, instructions under clauses 17.2 and 17.3 and possibly certain instructions empowered under clauses 8.3 and 8.4. There appears to be no reason in principle why he cannot also issue a compliance notice with all that implies under clause 4.1.2.

The architect cannot issue an instruction requiring a variation under clause 13, because the issue of the certificate of practical completion signifies, among other things, the end of the physical work. In *New Islington and Hackney Housing Association Ltd* v. *Pollard Thomas and Edwards* (2001), the construction work was let under two building contracts. Both contracts were on the IFC 84 Standard Form. In a comment that appears to be part of the *ratio*, Mr Justice Dyson said:

> 'On the true construction of the building contracts, [the architects] were authorised to issue variation instructions at any time up to practical completion of the works. But once practical completion had been achieved, the power to issue variation instructions was spent . . .'

2.11   *Provision of information*

Clause 5.2 requires the architect to provide the contractor with two copies of each of the contract documents, the contract drawings and the bills of quantities. This has to be done immediately after execution of the contract. That is to say, with all reasonable speed: *Hydraulic Engineering Co Ltd* v. *McHaffie, Goslet & Co* (1878). Since the proportion of building contracts actually executed before work begins on site must be very small indeed, strict adherence to this clause would see the contractor without these crucial documents, sometimes until after the project was complete. Of course, if the employer wanted the contractor to start work even though all the formalities were not finalised, he would have a strong interest in making sure that the architect issued the drawings, although technically too early.

As soon as possible after execution of the contract, clause 5.3.1.1 stipulates that the architect must provide the contractor with two copies of 'descriptive schedules or other like documents' which are necessary for the carrying out of the Works. The term 'descriptive schedule' is not defined and it is by no means easy to understand what is meant by 'other like documents'. One may surmise that a descriptive schedule is something like an ironmongery, lintel or door schedule. The only documents like those are other schedules. On the reasonable, but perhaps unjustified, basis that the draftsman would not have included the add-on phrase if he had not considered it necessary, 'other like documents' must add something; precisely what is unclear. Clause 5.3.2 demonstrates that the documents issued under clause 5.3.1.1 are purely explanatory of the contract documents and may not impose further or indeed different obligations on the contractor (this prohibition also extends to the master programme). What, then, is the difference between the information provided under clause 5.3.1.1 and what is to be provided under clauses 5.4.1 and 5.4.2?

Clause 5.4 was always a source of difficulty when it was part of JCT 80. Then, the problem was that although the architect's failure to provide information was a clear breach of contract under clause 5.4, the contractor was not entitled to, and the architect had no power to give, an extension of time unless the contractor had made specific written application for the information neither too early nor too late in relation to the time when it was needed. That deficiency could have been dealt with relatively easily by an amendment to the relevant event in question. Instead, a wholesale revision to clause 5.4 has taken place, together with a merciful shortening of clause 25.4.6.

Clause 5.4 is now in two parts. Clause 5.4.1 refers to the situation if there is an information release schedule. Clause 5.4.2 deals with the situation if there is no information release schedule or if the schedule does not include all the information needed by the contractor.

The sixth recital bluntly states that the employer has provided the contractor with a schedule stating what and when information will be released by the architect. Obviously someone thought it was a good idea to provide such a schedule when a moment's thought would have exposed the fallacy behind it. It is established that the contractor is entitled to construct the Works in any way he wishes provided only that he complies with the completion date and any sectional completion dates specified in the contract: *Wells* v. *Army & Navy Co-operative Society Ltd* (1902). In that case, Mr Justice Wright said:

> 'The plaintiffs must, within reasonable limits, be allowed to decide for themselves at what time they are to be supplied with details. The plaintiffs were entitled to do the work in what order they pleased.'

The same view was held by the trial judge whose judgment was affirmed by the Court of Appeal in *Greater London Council* v. *Cleveland Bridge & Engineering Co Ltd* (1986).

JCT 98 in clause 5.3.1.2 calls for the contractor to provide a master programme **[3.3]**. Invariably, this is provided by the contractor just before or at the commencement of work on site. Since the sixth recital speaks of the employer having 'provided' the schedule, the implication is that it has been provided at the time the contract is executed. We must assume for present purposes that a miracle has happened and the parties have executed the contract before the date for possession.

If the employer intends to provide the schedule, the contractor must be told when tenders are invited. Moreover, it is essential that each tenderer is provided with a copy of the schedule, because the contractor will need to know when he can expect the information so that he can plan the progress of the Works. Although the contract is silent on the point, it seems that the schedule must be prepared by the architect. He must prepare it without knowing how the contractor wishes to plan his work or whether he will find it more efficient or economical to start from north or south or even, at the time the schedule is prepared, who will be the contractor. So, where a schedule is to be used, the contractor will be deprived of his right to organise the work as he chooses.

An alternative is for the tenderers to be told that an information release schedule will be issued after it has been agreed between the architect and the successful contractor. A problem with that approach is the difficulty of reaching agreement and at the very time when, if the contractor is to be able to make a prompt start, the appropriate drawings should be available. A better alternative is to delete the sixth recital as hinted in footnote [e]. Anecdotal evidence suggests that in practice the provision of an information release schedule is something of a rarity.

Assuming that a schedule has been provided, the architect must provide the information listed on the schedule at the times stated. There are just two exceptions:

- If the contractor, by act or default, prevents such issue.
- If the architect and the contractor agree to alter the time for release. The clause stipulates that the agreement must not be delayed or withheld unreasonably. This suggests that the clause envisages that the architect may wish to delay an issue and wants the contractor to agree, or that the contractor wants the architect to agree to issue some information earlier. It is likely that the withholding of agreement in either circumstance could usually be justified as reasonable, depending on the precise circumstances.

Clause 5.4.2 is longer and, at first sight, appears to be similar to the old 5.4 of JCT 80. It shares some of the wording in the early part of the clause. The similarity ends there.

If there is no information release schedule or if it is not entirely comprehensive, clause 5.4.2 governs the release of other information. Even if a schedule is provided, it will never be comprehensive. The architect must do two things:

- Provide the contractor, from time to time as necessary, with two copies of the further drawings or details which are reasonably necessary to explain or amplify the contract drawings which have already been provided under clause 5.2.2; *and*
- Issue the instructions to enable the contractor to carry out and complete the Works in accordance with the contract.

On examination of the two things, the timing of the provision of the drawings and details is 'as and when from time to time may be necessary'. If this was not qualified by the remainder of the clause, it would suggest that the architect is obliged to keep up with the contractor's progress.

One of the terms of the contract is the date for completion. Therefore, if this was not also qualified by the remainder of the clause the architect's obligation would be to issue instructions to enable the contractor to complete by the date for completion.

The obligations so far as the timing is concerned are set out. For this purpose and for simplicity, reference to both drawings and instructions will be to 'information'. There are two situations:

- *The architect must supply information in accordance with the progress of the Works*
  That means that if the contractor is proceeding regularly and diligently with the Works, the architect must provide information to suit that progress. It also means that if the contractor is falling behind his programme and he is not going to finish by the completion date, the architect is entitled to adjust the supply of information accordingly provided the contractor receives the information when it is 'reasonably necessary' for him to do so. This clause allows the architect to issue information at a time when the contractor could not possibly use it to complete on time, even after the completion date has passed, provided the architect can say that he was issuing it to suit the contractor's actual progress on site. For the architect to take advantage of this power would be very ill-advised. If the contractor is late and if the architect does adjust his supply of information to match the contractor's progress, there is the real risk that at a subsequent date, the contractor may contend that the cause of his delay was the architect's failure to supply information to enable him to complete by the completion date. It is notoriously difficult in those circumstances to dispute the contractor's version of events, particularly in adjudication. To avoid that danger, architects should always endeavour to supply information at such times as will enable the contractor to complete by the date for completion even if realistically the contractor has no hope of achieving it.
- *The architect need not provide information to suit the contractor's progress* if, in the architect's opinion, the Works are likely to reach practical completion before the completion date. In those circumstances, his obligation is simply to provide the information so that the contractor can achieve the completion date. This provision is in accordance with the decision in *Glenlion Construction Ltd* v. *The Guinness Trust* (1987).

The big question is how the architect is supposed to know when the contractor needs to receive any particular information. In practice, the contractor will furnish his own schedules of information required and there is a powerful argument that the contractor is entitled to virtually all the information when the contract is executed. That is, indeed, what clause 5.3.1.1 may mean **[3.3]**. However, clause 5.4.2 sets out a procedure. If the contractor knows, but has reasonable grounds for believing that the architect does not know when the contractor should receive further information, the contractor must let the architect know. There are two qualifications:

- The contractor need only advise the architect to the extent that it is reasonably practicable. What that means is unclear. There is no requirement that the contractor must write to the architect, although it is always wise to put such matters in writing. Apparently it is sufficient if the contractor merely telephones the architect, tells him at a site meeting or just mentions it as the architect is carrying out a site inspection. It is difficult to envisage a situation when it will not be 'reasonably practicable' for the contractor to tell the architect that he needs some information.
- He must do so sufficiently in advance of when he needs the information so that the architect is able to provide the information in accordance with his obligations under this clause.

In practice, the problem would be that by the time the contractor realised that the architect was not going to provide the information, the latest time for receipt of the information would either be gone or so close that the architect would have no chance to meet the deadline. Contractors will no doubt continue to issue lists of all the information required and the latest dates required – a kind of 'information please release schedule'.

2.12   *The architect's obligations under JCT 98*

Clause 1.5 emphasises that it is wholly the contractor's responsibility to carry out and complete the Works in accordance with the contract. In this, it does no more than state the common law position, but it usefully sets out some of the common grounds on which contractors have been known to rely to excuse their own shortcomings. The clause makes clear that the following will not affect that obligation in any way:

- The architect's obligations to the employer
- The employer's appointment of a clerk of works
- Inspection visits to the Works or any workshop by the architect or the clerk of works
- Inclusion of the value of work, materials or goods in a certificate
- Issue of the certificates of practical completion or completion of making good defects.

For some reason, this clause is regularly overlooked. It has long been a principle that the architect's inspections on site are carried out for the benefit of the employer and the contractor cannot draw any conclusions from them. Certainly, the architect owes no duty to the contractor to find defects: *Oldschool* v. *Gleeson (Construction) Ltd* (1976). In *Bowmer & Kirkland Ltd* v. *Wilson Bowden Properties Ltd* (1996), Judge Bowsher said:

> 'Bowmer and Kirkland appear to be saying that the architects had a duty to supervise their work and maintain quality control, and if the architects failed to maintain quality control, Bowmer and Kirkland were to be excused from any defective performance of their duties under the contract. If that is their submission, it is wholly misconceived. The architects in this case were not under a duty to supervise, and even if they had been, their duty to supervise would have been owed to the employers, not to the builders, and if there had been a breach of a duty to supervise, that would not have excused the builders from maintaining their own system of quality control.'

In the light of such clear exposition of the position, what is to be made of clause 8.2.2? This clause first saw the light of day in one of the many JCT amendments to JCT 80. It says that if any materials, goods or workmanship included in the work are to be to the satisfaction of the architect under clause 2.1, the architect must 'express any dissatisfaction' within a reasonable time after the work is carried out. This seems to make the architect responsible for taking positive steps to find defects. He cannot, as hitherto, merely act as a bystander.

There are several points worth noting. The architect's obligation appears to be qualified by the reference to clause 2.1. That qualification is illusory. Clause 2.1 provides that if approval of materials or workmanship is something on which the architect should give his opinion, the materials and workmanship must be to his reasonable satisfaction. The provision was considered in *Crown Estates Commissioners* v. *John Mowlem & Co Ltd* (1994) together with the conclusivity provision in clause 30.9.1.1 of JCT 80. The Court of Appeal held that the clause should be read in a broad sense that the whole range of materials and workmanship were matters on which the architect should be satisfied. Therefore, clause 8.2.2 applies to all materials and all workmanship.

The clause does not say that the architect must express his dissatisfaction to anyone in particular. It might be argued that he can satisfy the clause simply by expressing his dissatisfaction to the employer. If that was the intention, the architect's obligation would better be stated in SFA/99 – the terms of engagement with his client.

For the clause to make sense, it must refer to the architect's obligation to express his dissatisfaction to the contractor.

It seems the expression need not be in writing. Clearly, the prudent architect will always record such things in writing. If he omits to do so, the stage is set for a dispute.

The question of what constitutes a 'reasonable time' in which the architect must act is always difficult to answer. It will depend on circumstances. It is suggested that the architect will comply with this requirement if he expresses his dissatisfaction before the contractor is ready to move to the next stage in the construction. Therefore, if the defect is several bad courses of brickwork, the contractor must be informed before he places further courses on top. However, architects do not usually visit site on a daily basis; therefore, his obligation must be further qualified with reference to the architect's first visit after the defective work is executed, by which time a metre height of brickwork could have been added. There will be some instances where a reasonable time will stretch almost to practical completion, for example, the laying of a concrete screed which is not due to receive any floor covering until the project is virtually complete.

If it is established that the architect did not comply with clause 8.2.2 in any particular instance, it is a breach of contract. There is no contractual machinery available to deal with it. If the employer is held to be vicariously liable, the contractor may be able to recover whatever damage he has suffered as a result. Thus, the employer may be put in the strange position of having to pay the contractor to dismantle and rebuild part of the structure in order to get at the defective item, because the architect has failed to spot the contractor's own breach of contract. Of course, he would not be liable for the correction of the original defect. The provision is ludicrous. The one saving grace is that there appear to be no grounds under which the contractor could obtain an extension of time to cover any resultant delay.

This is clearly a clause which is ripe for deletion.

2.13   Is the architect liable for breach of the Building Regulations?

Hitherto, it was established law that the only person liable for a breach of the Building Regulations was the contractor who actually carried out the work. The court expressly stated so in *Street & Another* v. *Sibbabridge Ltd* (1980):

'The builder is the person upon whom the regulations are binding. It is he who commits a criminal offence if the regulations are
breached.'

However, in *Blaenau Gwent Borough Council* v. *Khan* (1993), the Court
of Appeal held that an owner was the person 'carrying out the
work' within the meaning of Regulation 14(3) of SI 1985 1069,
because he authorised and commissioned the work. The court said
that the person carrying out the work could not be limited to the
person who physically did it. It seems, therefore, that the architect
who designs, or requires a contractor to execute, work in breach of
the Building Regulations may be prosecuted.

This is particularly pertinent in relation to clause 6.1.2 which
provides that if the contractor finds a divergence between statutory
requirements and any of the contract documents or architect's
instructions, he must immediately give a written notice to the
architect. The contractor's obligation is not to look for divergences,
only to report them if he finds any: *London Borough of Merton* v.
Stanley Hugh Leach Ltd (1985). Clause 6.1.3 allows the architect seven
days, after receipt of a notice or of his own discovery of a divergence, within which to issue an instruction and if the instruction
requires the Works to be varied, it is to be treated as a variation and
valued accordingly. Provided the contractor complies with his
obligation to notify the architect under clause 6.1.2 *if* he finds a
divergence, the contractor will not be liable to the employer if the
Works do not comply with statutory requirements and he has
simply complied with the contract documents and architect's
instructions (clause 6.1.5). Provided the contractor does not
blatantly ignore statutory requirements, it is unlikely that the
employer will be able to transfer liability to the contractor for any
failure of the Works to comply. In any event the architect would be
liable to the employer for any shortcomings in the construction
documents which he had produced as would the other consultants,
depending on their terms of engagement. The provisions of clause
6.1.5, however, would not protect the contractor against action by
the local authority for a failure to comply with the Building Regulations [3.5].

2.14 *The architect's design role under JCT 98*

The contract says nothing about the architect's role as designer.
However, it is clear that the contractor's obligation is purely to carry

out and complete the Works. Therefore, other than under clause 42 performance specified work **[3.13]**, someone other than the contractor is obliged to produce the information in the form of drawings and specification which comprise the Works which the contractor must carry out.

An architect has no authority apart from the express approval of the client to delegate any part of the design: *Moresk Cleaners* v. *Thomas Henwood Hicks* (1966) and in the absence of any terms or agreement to the contrary, the architect will be held liable for the whole of the design. However, if his conditions of engagement allow him to nominate and get the approval of the client for independent consultants, such as structural engineers or specialist subcontractors, to be appointed and paid by the client direct, he is not responsible for defects in their design. This is especially the case where there is an express clause that the client will hold the consultant and not the architect liable for the consultant's errors (see for example SFA/99 clause 3.11): *Investors in Industry* v. *South Bedfordshire District Council* (1985). But, if the independent specialist produces a design which any competent architect would know to be defective, the architect will be liable if he does not spot the defect and deal with it.

It will be noted that the professional man, unlike a supplier of goods or a contractor, gives no warranty as to fitness for purpose.

2.15   Duties of the quantity surveyor

Referring to JCT 63, the court in *County and District Properties Ltd* v. *John Laing Construction Ltd* (1982) described the quantity surveyor thus:

> 'His authority and function under the contract are confined to measuring and quantifying. The contract gives him authority, in certain instances to decide quantum. It does not in any instance give him authority to determine any liability, or liability to make any payment or allowance.'

The court was concerned whether a quantity surveyor had the authority under contract to make an agreement with a contractor which would be binding on the employer regarding the contractor's claims. The court was in no doubt that he did not have that authority. The same is largely true of JCT 98. For example, it is clear that although the quantity surveyor is to value variations and issue a valuation to the architect, if the architect considers them to be necessary, prior to the issue of a certificate (clause 30.1.2.1), the architect is responsible for the certification and he may override the valuation if he believes it to be incorrect.

The common practice of architects to simply transfer the figures from a valuation to a certificate is greatly to be deplored and such mindless reliance on the quantity surveyor probably amounts to negligence in certain circumstances. Clause 30.1.1.1 obliges the architect (not the quantity surveyor) to issue certificates and clause 30.2.1 makes clear that only work properly executed must be included. To safeguard his position, an architect should insist that the quantity surveyor provides a general breakdown of his valuation, sufficient to allow the architect to broadly satisfy himself that the valuation represents the work properly executed together with any further payments provided for in the contract. It is not suggested that the architect should attempt to redo the quantity surveyor's work or even minutely check every detail.

An exception to the quantity surveyor's limited role was introduced by JCT amendment 18 now largely incorporated (with mistakes corrected) into JCT 98. In clause 13.4.1.2 alternative A, it is the quantity surveyor and not the architect who effectively decides whether to accept the contractor's price statement, which may include loss and/or expense and the period of extension of time as well as the valuation of the instruction itself **[6.12]**. It is difficult to believe that the decision to give the quantity surveyor the final decision on these matters, against the tenor of the remainder of the contract, was anything other than inadvertent. Possibly the draftsman believed that if he required the quantity surveyor to come to his decision 'after consultation' with the architect, the problem would disappear. That would be to confuse 'consult' with 'agree'. The requirement to consult does not carry with it the implication to implement the result of the consultation. Therefore, the quantity surveyor may consult with the architect, the architect may state that the price statement should not be accepted and the quantity surveyor is entitled to accept it anyway.

It seems that if the quantity surveyor accepts the contractor's estimate of an extension of time for the instructed work, the architect is not required to confirm with any notice under clause 25. Indeed he must not do so, because clause 25 does not expressly provide for such a relevant event and clause 25.3.2 reinforces the position. It does this by providing that no decision of the architect can fix an earlier completion date than the completion date in the appendix, and then separately prohibits the fixing of the date earlier than the date accepted under clause 13.4.1. Presumably, although it is not stated in clause 25, the architect is expected to take into account the quantity surveyor's acceptance of an extension of time against the architect's wishes, when fixing any further completion dates as a result of subsequent delays. There are endless possibilities for problems if this situation arises.

2.16   *Duties of the clerk of works*

Clause 12 of JCT 98 carefully sets out the position of the clerk of works. It is clear that he is to be appointed by the employer. Occasionally, an architectural practice will have a permanent clerk of works on its staff and will use him as the clerk of works under clause 12. That is not a good idea so far as the architect is concerned. In *Kensington and Chelsea and Westminster Area Health Authority* v. *Wettern Composites* (1985) where the architects were held to be negligent, the damages were reduced by 20% because of the negligence of the clerk of works for whom the employer was vicariously liable on the basis of the usual liability of employers for employees. There would have been no reduction if the clerk of works had been employed by the architects.

A clerk of works must always be appointed where constant or frequent inspection is required. The contract emphasises that the clerk of works' only duty is to inspect. He is certainly not the architect's agent even though, in practice, architects often use the clerk of works to carry out many tasks such as checking dimensions on site and he is stated to be under the 'direction' of the architect for whom he is often also stated to be the 'eyes' and 'ears'. Presumably, so far as the contract is concerned, the only directions the architect may give must refer to the materials and operations which require inspection and the quality and standards specified.

The contractor is required to give all reasonable facilities to enable the clerk of works to inspect. That means that the clerk of works must be allowed access to the Works via scaffolding, but it does not mean that the contractor is required to erect scaffolding especially for the clerk of works. That would be to provide unreasonable facilities.

Clause 12 does not expressly empower the issue of directions by the clerk of works. Instead it seems to tolerate them. It simply says that if a direction is given, it is of no effect unless two criteria are satisfied:

- The direction is given in regard to something about which the architect is also empowered to issue an instruction; *and*
- The architect confirms the direction within two working days.

If confirmed, it becomes an architect's instruction effective from the date of confirmation. It is difficult to see why the provision regarding directions is included at all. The much simpler provision in IFC 98 is to be preferred. Indeed, it would be better if the contract stated that the clerk of works was not entitled to issue any directions or instructions, then there would be no confusion. At present, it is common for clerks of works virtually to run many jobs in the sense of answering questions from the contractor and even instructing variations. Many contracts would grind to a halt if that situation was not tolerated. Nevertheless, a good outcome is never an excuse for a bad practice and any contractor who acts on the basis of a clerk of works' instruction is in breach of contract.

Many clerks of works issue 'snagging lists', especially in the period before practical completion. They usually go much further than the usual list of defects which a clerk of works will hand to the contractor on a daily basis. There is, of course, no reference to 'snagging lists' in the contract although there is reference to a 'schedule of defects' in clause 17.2. 'Snagging' is properly used to denote a more informal kind of list. The contents still amount to breaches of contract on the part of the contractor. Clearly, clerks of works are trying to be helpful and their efforts are generally appreciated by contractors. Care should be taken, however, that the clerk of works is not simply doing the job which should be done by the contractor's person in charge. Another problem with snagging lists is that a contractor given such a list often assumes that the list is comprehensive and that if all the 'snags' are dealt with, the Works will be finished. The clerk of works seldom intends that to be the case and, even if he did, his intentions would be irrelevant. Merely complying with a clerk of works' snagging list can never take the place of compliance with the contract.

Some clerks of works insist on marking materials and goods which they consider not to be in accordance with the contract. There is no justification for that. Goods which are not in accordance with the contract are not accepted by the employer and, therefore, such goods remain the property of the contractor. As such, the employer or even the clerk of works personally could be liable for defacement.

CHAPTER THREE
THE CONTRACTOR'S OBLIGATIONS

3.1 Are there any implied terms in JCT 98?

The House of Lords considered this question in connection with the JCT 63 contract, where the obligations of the contractor were in similar terms to those in JCT 98, in *Gloucester County Council* v. *Richardson* (1967) shortly after their decision in *Young and Marten*. The case was concerned with whether a main contractor under JCT 63 was liable under an implied term for the fitness of materials supplied by a nominated supplier who had excluded the implied warranties of the then Sale of Goods Act 1893. The five Lords of Appeal were then unable to agree on whether these implied terms were to be read into JCT 63. Two thought they were; two thought they were not; and one did not express an opinion. As the learned editor of *Hudson's Building Law* remarked: 'That case poses more questions than it answers'.

But those were early days in the doctrine of implied terms in building contracts. There can be no doubt today that by operation of law any contractor under JCT 98 or the NSC/W or NSC/C contracts impliedly warrants that all that is done will be in a workmanlike manner, with materials of good quality (even if the materials are specified by the architect) and that these and the structure or any part of it will be reasonably fit for the purpose required if there is no independent designer involved [1.17].

3.2 Contractor's obligations under clauses 2 and 8

In addition to any obligations which the law will imply into the contract, clause 2.1 contains four separate obligations:

- To carry out and complete the Works in accordance with the contract documents
- In doing so to use materials of the quality and standards specified therein

- In doing so to use workmanship of the quality and standards specified therein.
- There is a proviso that where and to the extent that the approval of quality or standards is a matter for the opinion of the architect, they will be to his reasonable satisfaction.

The last obligation was considered by the Court of Appeal in *Crown Estates Commissioners* v. *John Mowlem & Co* (1994) **[5.16]**.

Does the contractor have dual obligation to comply with the specification *and* to reasonably satisfy the architect? The question has been concerned in a number of cases, notably *National Coal Board* v. *Neill* (1984) where it was held that it is a matter of interpretation in each case whether the contractor has a dual obligation. It seems that generally, if the obligation is clearly linked with the word 'and', the contractor will have a dual obligation, so that even if the architect expresses his reasonable satisfaction in regard to some matter, it will not protect the contractor if it can be demonstrated that he has not complied with the contract documents: *Billyack* v. *Leyland Construction Co Ltd* (1968). That certainly seems to be the case on a strict reading of the wording in clause 2.1. However, if the architect has expressed his reasonable satisfaction to specific items of materials or workmanship by the issue of a final certificate, the position is more complex.

The conclusive effect of the final certificate as expressed in clause 30.9.1.1 was amended in the JCT 80 edition by JCT Amendment 15 following the *Crown Estates Commissioners* case which held that the previous wording was to be given a wide meaning so that essentially all instances of quality and workmanship were matters for the opinion of the architect. The amended clause, now incorporated into JCT 98, modifies the previous position. Now, the final certificate is only conclusive evidence of the architect's reasonable satisfaction in respect of materials or workmanship which are specifically stated in either the contract drawings or bills or in any numbered document or in any architect's instruction or in any further drawings issued under the terms of the contract to be to the architect's reasonable satisfaction. It is likely that conclusivity will not be triggered in these instances unless the document in question refers to the specific qualities or standards which are to satisfy the architect. That seems clear from the use of the words 'particular qualities' and again later 'particular standard'. Therefore, on a strict reading of the clause, if the architect refers to 'all plaster' to be to his approval, it may be argued that it would not trigger conclusivity. It seems the architect would have to be more precise, for example by referring to 'the smoothness of all plaster surfaces'. The position is by no means clear and architects can avoid disputes in this area by adopting either of two expedients:

(1) Being careful not to reserve anything for his approval or satisfaction; *or*
(2) After reserving anything for his approval, to be assiduous in inspecting the relevant materials or workmanship.

In general, architects will be wary of reserving anything for approval. Where there is such an exceptional term in the contract documents, the contractor has no option but to comply with it. The architect does not have to exercise his powers under clause 8.4.1 to require the removal of any materials or workmanship which do not meet with his approval in these circumstances. It is sufficient if he indicates that they are not to his satisfaction. Likewise, he is under no obligation to issue a variation order to secure replacement of materials of which he disapproves. Neither is he under an obligation to include in interim certificates work which is not to his satisfaction – quite the contrary.

The contractor's obligations under clause 2.1 are amplified by clause 23.1 which obliges him, on the date of possession, to 'thereupon begin the Works'. He must regularly and diligently proceed with them so as to complete on or before the completion date, that is the date stated in the contract or any extended date. The Court of Appeal in *West Faulkner* v. *London Borough of Newham* (1995) have helpfully defined 'regularly and diligently':

> 'What particularly is supplied by the word "regularly" is not least a requirement to attend for work on a regular daily basis with sufficient in the way of men, materials and plant to have the physical capacity to progress the works substantially in accordance with the contractual obligations.
>
> What in particular the word "diligently" contributes to the concept is the need to apply that physical capacity industriously and efficiently towards the same end.
>
> Taken together the obligation upon the contractor is essentially to proceed continuously, industriously and efficiently with appropriate physical resources so as to progress the works steadily towards completion substantially in accordance with the contractual requirements as to time, sequence and quality of work.'

The contractor can have no claim of any kind for expense or loss caused by the removal and replacement of materials or workmanship which were to be, but were not, to the architect's satisfaction for it is the contractor who is in breach.

Clauses 8.1.1 and 8.1.2 are partly unnecessary and repetitious, because they provide that materials, goods and workmanship must be of the kinds and standards in the contract bills. That has already been stipulated by clause 2.1 and it would in any event be necessarily implied. However, clause 8.1.1 introduces an inconsistency which may be important. Materials and goods have only to be provided 'so far as procurable' whereas the contractor's obligation under clause 2.1 is to provide what is specified in the contract documents. However, the introduction in clause 8.1.1 of the word 'procurable' affords the contractor a valuable protection if materials or goods are truly unobtainable. There is no protection for a contractor who finds it more difficult or expensive to provide what is specified. Taken literally, the contractor is protected even if the materials or goods were not procurable when the contract was executed. A sensible and businesslike construction of the provision would confine the meaning of 'procurable' to those items whose status has changed since the contract was executed. However, it is not at all certain that the provision should be interpreted in that way. There is no authority on the matter and although the author has previously leaned towards this sensible and businesslike interpretation, a strict reading of the clause results in the harsh conclusion, so far as the employer is concerned, that if the items are not procurable for any reason, the contractor's obligation to provide them is at an end. It then becomes necessary for the architect to issue an architect's instruction requiring as a variation the provision of a substitute material. The variation is to be valued in the usual way. This conclusion has the effect of removing from the contractor any obligation to check that specified goods and materials are procurable before tendering. In order to moderate this harsh view and return to a sensible view, it may be necessary to substitute some such words as 'except in so far as they become not procurable after the Base Date'.

Both clauses 8.1.1 and 8.1.2 include the contractor's obligation to comply with the contractor's statement issued in regard to any performance specified work, and go on to relate the contractor's obligation to the obligation to provide goods, materials and workmanship to the architect's reasonable satisfaction as set out in clause 2.1.

Clause 8.1.4 stipulates that the contractor may not substitute

goods or materials for goods or materials in the contractor's statement under clause 42.2 unless the architect has given consent. The
consent may not be unreasonably delayed or withheld, but it is
expressly stated that such consent does not relieve the contractor of
any contractual obligations. This may be a 'catch-all' proviso, but
two immediate results are:

- The contractor's obligation to satisfy the original performance
  obligation remains intact. So that if the substitution of a material
  proposed by the contractor and given consent by the architect
  results in a failure by the contractor to satisfy the performance
  specification, the contractor will be liable for such failure. This
  straightforward position is complicated by the provision of
  clause 42.17.1.2 which prevents the contractor from assuming
  liability for fitness for purpose **[3.13]**.
- The substitution is not to be treated as a variation under clause
  13.

The second part of clause 8.1.2 elaborates on the contractor's basic
obligations under clause 2.1 to say that to the extent that no standards are specified, the standard must be 'appropriate to the
Works'. This strongly suggests that if nothing is properly specified,
the contractor has the task of deciding what is appropriate. Can the
contractor simply decline to choose and request an instruction
stating what is required? The contractor must be able to do that
because, except for performance specified work and the possible
inclusion of the Contractor's Designed Portion Supplement, he has
no design responsibility under JCT 98. However, if the contractor
proceeds to use a standard of workmanship he believes is appropriate, he may be liable if it is not in fact appropriate. What is or is
not appropriate in any particular instance may be referred to an
adjudicator **[11.3]**. It is useful to consider this provision in relation to
the decision of the Court of Appeal in *Rotherham Metropolitan
Borough Council* v. *Frank Haslam Milan & Co Ltd and M.J. Gleeson
(Northern) Ltd* (1996) which refused to place on the contractor the
obligation of ensuring fitness for purpose of fill material where the
architect had given a general description 'slag' and the contractor
had used a material which complied with the description, but
which was nevertheless unsuitable. The court held that the
employer was relying on the expertise of the architect and not the
contractor. The contractor had strictly complied. There were particular differences between this case and the situation envisaged by
clause 8.1.2. In the first place the contract used in the *Rotherham* case

was the 1963 edition (1977 revision) of the JCT Standard Form of Contract. Secondly, the case referred to a material and not a standard of workmanship. Finally and in any event, the architect in the *Rotherham* case had not failed to describe the material. He had, in fact, described several materials which the contractor could use. It was not a good specification because, by being too broad, it effectively allowed the contractor to use an unsuitable material.

Clause 8.1.3 requires the contractor to carry out all work in a 'proper and workmanlike manner' and according to the health and safety plan. When this provision was introduced into JCT 80, guidance notes were provided. Such notes are not of course contract provisions and whatever may have been the intention of the draftsman it is only the intentions of the parties, gleaned from the words in the contract, which can be taken into account. There is clearly a difference between 'workmanship' and carrying out work 'in a workmanlike manner'. The reference to the health and safety plan also provides a clue. This clause is not concerned with the permanent Works, but with the way in which the finished product is achieved. To take a simple example: a steel beam is obviously material or goods; its final position, centralised on padstones and truly horizontal, and the finish of the paint applied to it are all matters of workmanship. However, the act of lifting and manoeuvring the beam into position and settling it on the padstones may or may not be accomplished in a workmanlike manner. In addition to the strictly constructional aspects of the task, there are questions of safety in handling for the operatives engaged on the task and for any person or property which might be injured or damaged during the process.

A key case in this regard was *Greater Nottingham Co-operative Society* v. *Cementation Piling and Foundations Ltd* (1988) which triggered the introduction of clause 8.1.3. The case revolved around whether the piling sub-contractor had a duty to carry out the work with reasonable skill and care. It was held at first instance, and not the subject of the appeal, that there was no such duty express or implied.

Some contractors are unaware that, under clause 8.2.1, they are obliged to provide the architect, on request, with documents (the contract refers to 'vouchers') to prove that the goods and materials comply with the requirements of clause 8.1. Of this clause only subsections 8.1.1 and 8.1.4 appear to be relevant. Although the contractor must 'prove', the standard of proof is only 'the balance of probabilities'.

The remainder of clause 8 has been considered in **[2.5]**.

3.3　Contractor's master programme and other documents

Many architects believe that the contractor is obliged to provide them with a master programme and, moreover, to update the programme on request. This is a misconception. The misconception goes further to the extent that an architect may think that a contractor has an obligation to comply with his own programme. All that may make very good sense, but the reality is quite different.

Clause 5.3.1.2 provides that the contractor must provide the architect with a copy of his master programme. Nothing obliges the contractor to have a master programme. Therefore, if he does not have one, he can have no obligation to provide it. This clause has exactly the same structure as clause 5.3.1.1 which requires the architect to provide the contractor with descriptive schedules and the like. It does not require the architect to produce the schedules. If the architect believes they are not required in any particular instance, he may opt not to produce them. If the architect does not produce the schedules, he has no obligation to provide them **[2.11]**.

Although it is probably unlikely that the contractor will not have a programme on even the smallest project, it may be worthwhile amending the clause to oblige the contractor to produce a master programme showing his intentions. There is nothing to prevent the architect including in the specification or the preliminaries to the contract bills, more details of the kind of programme required. That kind of requirement does not fall foul of the priority clause (2.2.1), because it does not attempt to 'override or modify' what is in the contract, but merely to add to it. Among other things, it is useful if the contract bills require the programme to be in network or precedence diagram form as well as a bar chart, fully resourced and indicating all the logic links. The use of such a programme when analysing such matters as extensions of time has been approved by the court: *John Barker Construction Ltd* v. *London Portman Hotels Ltd* (1996) **[7.16]**.

The contract neither requires nor empowers the architect to make any comment about the contractor's programme, much less to approve it. If the architect, on receipt of the programme, notices anything which causes him concern, it makes sense to convey this concern to the contractor, but it cannot be done in the form of an instruction. It is best done as a question, e.g. 'Are you sure that you have allowed sufficient time/labour/plant to complete the foundations within four weeks?'. Even if the architect does approve the contractor's programme, it is doubtful whether it has any significance: *Hampshire County Council* v. *Stanley Hugh Leach Ltd* (1991) unless the programme shows an early completion date (see below). Where the architect does approve the programme, he is probably doing no more than signifying in a broad way that if the contractor carries out the work in accordance with the programme, the architect will be satisfied with the progress. Approval does not transfer responsibility for the contents of the programme to the architect. Whether or not it is correctly calculated taking all resources into account is always a matter for the contractor.

There is no contractual provision which obliges a contractor to comply with the programme he has submitted. In practice, it would be very perverse of a contractor to proceed to execute the Works in a completely difference sequence as well as being highly unlikely, but he can do so if he wishes. It follows, therefore, that individual dates in a programme have no binding effect. The programme is not a contract document, although it may perhaps be termed a 'contractual document', being generated by a term in the contract. It would be possible, although rarely advisable, to make the programme a contract document. The result would be that every deviation potentially would be a breach of contract on the part of the author of the deviation: *Yorkshire Water Authority* v. *Sir Alfred McAlpine and Son (Northern) Ltd* (1985) **[6.2]**.

So far as updating the programme is concerned, clause 5.3.1.2 requires the contractor to provide an amended programme within 14 days if:

- The architect makes a decision about an extension of time under clause 25.3.1; *or*
- There is a confirmed acceptance of a 13A quotation.

Strangely, there is no reference to the acceptance of a contractor's price statement under clause 13.4.1.2 alternative A which may also have the effect of extending the time for completion.

There will be many instances when an architect will want the contractor to provide an updated programme although he has neither given an extension of time nor confirmed acceptance of a clause 13A quotation. A common example is when a contractor's progress is seriously delayed due to his own fault. The answer is to amend the clause to allow the architect to require the contractor to provide an updated programme upon the architect's reasonable request.

It is quite common for a contractor to produce a programme which shows that he intends to complete the Works before the date for completion stated in the appendix. There is nothing to prevent him doing so, because clause 22.1 requires him to complete the Works 'on or before' the completion date. However the architect is not obliged to provide information at the right times to enable the contractor to complete early: *Glenlion Construction Ltd* v. *The Guinness Trust Ltd* (1987) **[2.11]**. The submission of a programme showing early completion may be taken into account by the architect when considering whether an extension of time is due in any particular circumstance.

The contractor must keep on site a copy of the contract drawings, a copy of the unpriced bill of quantities, a copy of descriptive schedules 'or other like documents', a copy of his master programme (if he has one) and a copy of further drawings (clause 5.5).

The contract provides that the contractor must give to the employer or to the architect various notices. The mandatory notices are listed in Table 3.1.

Table 3.1
Mandatory notices from the contractor

Clause 2.3	Immediately in writing, upon discovery of any discrepancies in or between the drawings, the bills of quantities, architect's instructions and numbered documents. If the contractor finds any discrepancies in or between the contract documents, etc.
Clause 2.4.1	If the contractor finds any discrepancies between his performance specified work statement and an architect's instruction.
Clause 5.3.1.2	If he has a master programme, two copies to the architect and of every subsequent amendment.
Clause 5.4.2	Requesting information, if the contractor is aware and has reasonable grounds to believe that the architect is not aware that information is required by a specific time.
Clause 6.1.1	As required by Act of Parliament, regulation, or byelaw.
Clause 6.1.2	Immediately in writing, any divergence he has discovered between statutory requirements and any of the documents noted in clause 2.3.
Clause 6.1.4.2	Forthwith of emergency compliance with statutory requirements.
Clause 6A.2	Notify the employer if the contractor makes any amendment to the health and safety plan.

Table 3.1 *Contd*	
Clause 13.2.3	In writing within seven days of receipt of a 13A instruction, if the contractor disagrees with the application of clause 13A to an instruction.
Clause 22A.4.1	Forthwith in writing (and to the employer), notice of the extent, nature and location of damage caused by any of the all risks to the work executed or site materials.
Clause 22B.3.1	Forthwith in writing (and to the employer), notice of the extent, nature and location of damage caused by any of the all risks to the work executed or site materials.
Clause 22C.4	Forthwith in writing (and to the employer), notice of the extent, nature and location of damage caused by any of the all risks to the work executed or site materials.
Clause 23.3.3	Notify the employer of the amount of any additional premium.
Clause 25.2.1.1	Forthwith if and whenever it becomes reasonably apparent (with a copy to any nominated sub-contractor mentioned), notice that the progress of the Works is likely to be delayed stating material circumstances, the cause and identifying any relevant events.
Clause 25.2.2	As soon as possible after a clause 25.2.2 notice identifying a relevant event, give separate particulars of the effects of the delay and the estimated effect on completion date of every identified relevant event.
Clause 25.2.3	Give such further notices in writing as reasonably necessary or requested by the architect to update previous notices under clause 25.
Clause 26.1.2	Upon the architect's request, submit the information as will reasonably enable the architect to form an opinion.
Clause 26.1.3	Upon the architect's request, submit the information as is reasonably necessary for ascertainment.
Clause 26.4.1	Upon receipt of written application by a nominated sub-contractor for loss and/or expense under NSC/C.
Clause 27.3.2	Notify the employer if the contractor has made a composition or arrangement with creditors or, if a company, has made a proposal for voluntary arrangement.
Clause 28.2.1	If the employer has defaulted under clause 28.2.1 and the contractor wishes to set the determination process in motion.

Table 3.1 *Contd*

Clause 28.2.2	If the employer has defaulted under clause 28.2.2 and the contractor wishes to set the determination process in motion.
Clause 28.2.3	If the contractor wishes to determine his employment subsequent to the clause 28.2.1 and 28.2.2 notices.
Clause 28.3.3	If the contractor wishes to determine his employment subsequent to the employer's insolvency.
Clause 28A.1.1	If the contractor wishes to determine his employment subsequent to the occurrence of events under this clause.
Clause 30.6.1.1	Not later than six months after practical completion, all documents necessary for adjustment of the contract sum including nominated sub-contractor and supplier information.
Clause 31.5.1.1	A statement to the employer showing the direct cost of materials, if the authorisation is a CIS 4 registration card.
Clause 31.7	If the authorisation is a CIS 4, replaced by CIS 5 or CIS 6, inform the employer.
Clause 31.8	If CIS 5 or CIS 6 is withdrawn, inform the employer.
Clause 34.1.3	Inform the architect or clerk of works on discovery of a fossil.
Clause 35.8	In writing if unable within 10 working days to reach agreement with a proposed nominated sub-contractor.
Clause 35.13.3	Before the issue of each interim certificate (other than the first), provide reasonable proof of discharge of payments to nominated sub-contractors.
Clause 35.15.1	If the contractor requires the architect to issue a certificate of non-completion of the nominated sub-contract works.
Clause 35.24.6.2	Informing the architect if the employment of the nominated sub-contractor has been determined.
Clause 42.15	Within seven days of receipt of the relevant instruction if the contractor considers that it injuriously affects the efficacy of the performance specified work.

3.4   *Statutory obligations*

JCT 98 clause 6.1 imposes on the contractor an obligation to comply with all statutory obligations. It also imposes an obligation on him by clause 6.1.2 to give immediate notice to the architect in writing if he discovers any divergence between the contract documents, numbered documents, drawings and architect's instructions and statutory obligations which include, of course, the Building Regulations.

This seems to cover, among other things, the situation where the contractor finds out that, because of soil conditions, the designed foundations will not comply with the Building Regulations. That part of clause 6 ends with a proviso in clause 6.1.5 that if the contractor complies with his obligation to notify the architect if he discovers a divergence and otherwise the contractor constructs the Works in accordance with the contract documents and any further drawings and architect's instructions, he will not be liable to the employer if the Works do not comply with statutory requirements as a result. In short, the clause purports to exonerate the contractor from any liability to the employer if he builds the architect's design **[4.14]**.

It is likely that this clause may not always be effective in achieving its intended purpose if the employer can be described as a 'consumer'.

Clause 6.1.5 is clearly caught by section 7 of the Unfair Contract Terms Act 1977. This applies to situations 'where the possession or ownership of goods passes under or in pursuance of a contract not governed by the law of sale of goods or hire purchase ...'. That is, it applies to construction contracts. It applies to 'contract terms excluding or restricting liability for breach of obligation arising by implication of law from the nature of the contract'.

Section 7.2 then provides that 'as against a person dealing as a consumer, liability in respect of goods ... quality or fitness for any particular purpose cannot be excluded by reference to any such term', and only in other circumstances, if reasonable. The primary liability, both criminally and civil for breach of the Building Regulations, rests with the contractor, and he will be liable to the employer if he fails to comply with the Regulations. Therefore, he can only escape from his liability if JCT 98 clause 6.1.5 is reasonable. The contractor may have recourse in tort for an indemnity for his loss from the architect, both at common law and under the Civil Liability (Contribution) Act 1978 or for a contribution under that Act. Quite apart from legal considerations, this must be common sense **[2.13]**.

3.5   *Implied terms: Building Regulations*

In *Street and Another* v. *Sibbabridge Ltd and Another* (1980) Judge Fay QC held that it was an implied term of all building contracts that the contractor would comply with the Building Regulations. This obligation overrode the architect's design and instructions.

In that case, the defendant contractor had complied exactly with the designer's instructions as to the depth of the foundations for a garage. But due to the presence of trees in the vicinity, the foundations proved inadequate and, therefore, in breach of the Building Regulations. The judge said:

'The drawing provided for foundations at a depth of 2 ft 9 in, a concrete strip under the walls and that is what the builder put in.

It is said that there were implied terms in the contract: firstly that [the builder] would construct the garage in accordance with the Building Regulations which provide that "the foundation of the building shall be taken down to such a depth ... and so constructed as to safeguard the building against damage by swelling or shrinking of the sub-soil".

And secondly, "that the [builders] would use reasonable care and skill in constructing the said foundation".

These are almost common form implied terms of building contracts and I find these implied terms to be present in this case.

The term that the building would be constructed in accordance with the Building Regulations must flow from the fact that the builder is the person upon whom the Regulations are binding. It is he who commits a criminal offence if the Regulations are breached.

I have had an interesting argument addressed to me by [counsel for the builders] upon the curious situation which arises where there is an express term conveyed by the incorporation, as it undoubtedly was into this contract, of the drawings ... that had a foundation 2 ft 9 ins deep ... contrasted with an implied term which ... would require another type of foundation, either deeper or different in character.

In the circumstances of this case, the express term cannot be said to prevail over the implied term in any sense.

The obligation to comply with the Regulations is in terms absolute. The implied undertaking to comply with the Regulations must override any matter in the plans incorporated into the contract which it conflicts with.

That is only common sense, and there is evidence that it is also

common practice, because the architect called by the [builder]
said in his report: "It is established convention in the building
industry that drawings of foundations are provisional only and
are subject to review on excavation and to remeasurement on
completion".

   If there is such a conflict, then the implied term must prevail.'

This is an important case. It makes clear that implied terms which in
the words of section 7 of the Unfair Contract Terms Act 1977 arise
'by implication of law' override express terms of the contract to the
contrary **[1.17]**. By contrast, terms which might be implied to give
commercial effectiveness to a contract under *The Moorcock* doctrine
cannot override the contract if there is an express term which covers
the same subject matter: *Les Affréteurs Réunis* v. *Leopold Walford*
(1919).

   However, many people still believe that there cannot be an
implied term if there is an express term covering the same subject
matter. That is a fallacy in the case of implied terms 'arising by
implication of law'. In *Young and Marten* v. *McManus Childs Ltd*
(1968) the roofing sub-contractor had complied exactly with the
contractual specification; nevertheless, the House of Lords held that
he was liable for breach of a term implied by operation of law.

   In the *Street* case, the judge also held that there was an implied
term that the builder 'would use reasonable care and skill in constructing the foundations':

> 'Although he honestly believed that his foundations were good
> ones, he had followed the plans, the foundations had been
> approved by the building inspector and were well constructed,
> he departed from the duty that the law lays upon him ... by
> failing to measure up to the state of knowledge which a builder
> ought to have.'

There are two questions which seem to be important:

- Did a contractor in a particular case know that he would not
  comply with the Building Regulations by following the architect's design?
- If the contractor did not know, was it something which he should
  have known, as an averagely competent contractor?

Although *Street* answered the last question in the affirmative, more
recent cases may not always come to that conclusion – see *Rotherham*

Metropolitan Borough Council v. *Frank Haslam Milan & Co Ltd and M.J. Gleeson (Northern) Ltd* (1996).

3.6   Person-in-charge

Clause 10 provides that the contractor must constantly keep a competent person-in-charge on the Works. There is no requirement, as in some forms of contract, that such a person should be named by the contractor or agreed by the architect. The expression 'foreman-in-charge' which was used at one time has been changed to what the JCT describe as 'the more neutral phrase'. Presumably this is to allow for the circumstance in which the person-in-charge may be a woman. Some contracts use the term 'site manager' or 'site agent'.

The word 'constantly' is not to be taken literally. A better view is that the person-in-charge must be constantly on the Works during any period when work is in progress.

Instructions given by the architect or directions given by the clerk of works are deemed to have been given to the contractor. This is particularly important where the architect hands written instructions to the person-in-charge on site and does not confirm them to the contractor's head office.

3.7   To obey architect's instructions

The contractor has an obligation under clause 4.1.1 to obey all instructions given by the architect which are within his powers under the contract. The contractor is entitled to be provided with details in writing, stating the clause of the contract under which the instructions are given. The contractor may request this information from the architect (clause 4.2).

The architect must specify a clause and the contractor can choose whether or not to dispute it. If he chooses to dispute it, the matter can be resolved by adjudication or whichever of the alternative dispute resolution procedures, arbitration or legal proceedings, are stipulated in the contract. If the contractor chooses not to dispute the specified clause, the instruction is deemed empowered by the clause for all the purposes of the contract. This means that the contractor will be estopped from subsequently claiming that the architect's instruction was not validly given. In view of the clear power to challenge given in clause 4.2, the result is probably the same if the architect does not specify the empowering clause and the contractor does not exercise his right to request the information.

A very important consideration is sometimes overlooked. If the contractor does not query the clause and, therefore, the clause is deemed to be empowered 'for all the purposes of this Contract', one of the 'purposes' is to enable the instruction to be valued if appropriate. The contractor must not comply with instructions which are not empowered and if he does so, he is in breach of contract and certainly not entitled to payment. Clause 4.2, therefore, provides an important safeguard for the contractor.

Under clause 4.1.1, the contractor also has the right to object to certain instructions. These are instructions which amount to variations under clause 13.1.2 **[6.2]** dealing with the imposition or alteration of obligations or restrictions in respect of access to, or use of, any part of the site, limitation of working space or hours and the execution of the work in a particular order.

The contractor is not confined to objecting; he may refuse to carry out such variations to the extent that his objection is reasonable and made in writing. Use of the phrase 'to the extent that' suggests that if the contractor's objection is directed only at part of the variation, he must comply with the other part. How exactly that will be achieved will depend on circumstances, but it is likely to throw up problems. If there is dispute about whether the objection is reasonable, either party may refer the matter to adjudication. One sympathises with an adjudicator in this situation since there are no guidelines for him to follow and he will be left to such native instincts as he may possess.

There is an express sanction in clause 4.1.2 which the architect can use if the contractor fails to comply with his instruction. The architect may issue a written notice to the contractor requiring compliance with the instruction. If the contractor has not complied within seven days after receipt of the notice, the employer may engage other persons to do whatever is necessary to carry out the instruction. It is good practice, although not essential, for the architect to issue a further notice to the contractor at the end of the seven day period, setting out what the employer intends to do and pointing out that any last minute attention to the instruction on the part of the contractor will not alter the contractor's liability to the employer for any irrecoverable costs.

Wherever practicable, the employer should invite three other persons to tender for the work in order to mitigate his loss and to counter any future contentions that the work was needlessly expensive. The employer is entitled to deduct all costs incurred *in connection* with the exercise from money due to the contractor (after serving the proper written notices [5.21]) or to recover them as a debt. Use of the italicised phrase broadens the scope of such costs to include all additional money the employer must expend over what he would have had to pay the original contractor: *Ashville Investments Ltd* v. *Elmer Contractors Ltd* (1987). Clearly, an extra architect's and quantity surveyor's fees necessarily incurred are recoverable as part of the total additional costs.

It has been suggested that if the architect and employer fail to take prompt action on the contractor's failure to comply, the employer may be taken to have waived his rights to take action subsequently. Such an outcome seems unlikely.

3.8   Setting out

Clause 7 obliges the architect to decide on levels and to provide the contractor with accurately dimensioned drawings showing the information necessary to enable the contractor to set out the Works at ground level. This is a strange requirement, because ground level may not be the most appropriate level at which to set out the building in some circumstances. Although it might be expected that the architect will provide the most appropriate information to suit a particular building on a particular site, strictly he will be in breach of his obligation if he does not show the setting out at ground level. On the other hand, the contractor cannot complain if he receives ground level setting out even though it may be quite inappropriate in a particular instance. However, he would be entitled to additional information under the provisions of clause 5.4.2 [2.11].

The obligation to provide 'accurately dimensioned drawings' echoes the words of Mr Justice Vinelott in *London Borough of Merton* v. *Stanley Hugh Leach Ltd* (1985) that the architect's general obligation was to provide correct information. The contractor is entitled to rely on the accuracy of the setting out information, subject to any implied obligation to warn the employer if he becomes aware of a serious error [3.14]. This probably amounts to an indemnity for the contractor in the case of any action for trespass resulting from inaccurate information.

The contractor has an obligation to set out accurately and if the setting out is inaccurate, he must rectify it at his own cost. If the discovery is made at a late stage, rectification may involve the expenditure of many thousands of pounds on the part of the contractor. The law is perfectly clear. In such a case, the employer is entitled to be put in the same position, so far as money can do it, as if the contractor had set out correctly: *Robinson* v. *Harman* (1848). This strict view is apt to be modified in practice if the benefit gained by the employer is substantially outweighed by the cost of rectification: *Ruxley Electronics and Construction Ltd* v. *Forsyth* (1995). In any event, it is open to the architect, if the employer consents, to instruct the contractor not to amend the error. The clause then states that an 'appropriate deduction' is to be made to the contract sum. 'Appropriate deduction' in these circumstances will not be the same as is referred to in clause 17.2 **[12.4]**. How the deduction is to be calculated is a matter of conjecture. It will clearly depend on the effect of the error and perhaps any continuing expense to the employer, e.g. the increased cost of heating a larger than expected building.

3.9   Prohibition against assignment

JCT 98 contains the familiar prohibition, in clause 19.1.1, against assignment of the contract. Both parties are prohibited. The words, as they stand, are misleading. Nobody can ever assign a contract; all they can ever do is to assign the benefits of a contract – never the obligations. The assignment of the benefits of a contract is the assignment of a chose-in-action and is governed by section 136 of the Law of Property Act 1925.

The real meaning of these words, in the light of their ancestry, is that they are intended as a prohibition against vicarious performance of the obligations of the contract and must be so construed. Originally, 'neither in law nor in equity could the burden of a contract be shifted off the shoulders of a contractor on to those of another, without the consent of the contractee'. Later, the common law would allow vicarious performance, that is, performance by another person, without the consent of the other party, but only where the obligations were not personal. Even where clothes are sent to cleaners, the Court of Appeal held in *Davies* v. *Collins* (1945) that this was a personal obligation which could not be subcontracted out without it being a breach of contract. But the repair of railway waggons, with 'a rough description of work which ordinary workmen conversant with the business would be perfectly able to execute', was held to be one that could adequately be performed by somebody other than the party who contracted to do it: *British Wagon Co* v. *Lea* (1880).

Generally speaking in most building work, in the absence of custom or contract to the contrary, vicarious performance will be unlawful. The common practice of sub-contractors sub-sub-contracting out part of their work without consent is normally a breach of their contract. It is this which the clause is intended to prohibit, even though clause 19.2.2 deals with the same subject matter, preventing the contractor from sub-letting any part of the Works without the written consent of the architect who is not to unreasonably delay or withhold it. The second part of the clause makes clear that sub-letting does not relieve the contractor of any responsibility for carrying out the Works in accordance with clause 2.1.

Where the architect consents to sub-letting, the sub-contractor is termed domestic. The form intended to be used for the sub-contract is DOM/1. Clause 19.2.1 usefully states that, unless he is a nominated sub-contractor, any person to whom the contractor sub-lets any part of the Works is termed a 'Domestic Sub-Contractor'.

The difference between assignment, sub-contracting and novation was set out with admirable clarity by Staughton LJ in *St Martins Property Corporation Ltd and St Martins Property Investments Ltd* v. *Sir Robert McAlpine & Sons Ltd and Linden Gardens Trust Ltd* v. *Lenesta Sludge Disposals Ltd, McLaughlin & Harvey PLC, and Ashwell Construction Company Ltd* (1992) in the Court of Appeal:

'(a) Novation This is the process by which a contract between A and B is transformed into a contract between A and C. It can only be achieved by agreement between all three of them, A, B and C. Unless there is such an agreement, and therefore a novation, neither A nor B can rid himself of any obligation which he owes to the other under the contract. This is commonly expressed in the proposition that the burden of the contract cannot be assigned, unilaterally. If A is entitled to look to B for payment under the contract, he cannot be compelled to look to C instead, unless there is a novation. Otherwise B remains liable, even if he has assigned his rights under the contract to C...

(b) Assignment This consists in the transfer from B to C of the benefit of one or more obligations that A owes to B. These may be obligations to pay money, or to perform other contractual promises, or to pay damages for a breach of contract, subject of course to the common law prohibition on the assignment of a bare course of action. But the nature and content of the obligation, as I have said, may not be changed by an assignment. It is this concept which lies, in my view, behind the doctrine that personal contracts are not assignable... Thus if A agrees to serve B as chauffeur, gardener or valet, his obligation cannot by an assignment make him liable to serve C, who may have different tastes in cars, or plants, or the care of his clothes ...

(c) Sub-contracting I turn now to the topic of sub-contracting, or what has been called in this and other cases vicarious performance. In many types of contract it is immaterial whether a party performs his obligations personally, or by somebody else. Thus a contract to sell soya beans, by shipping them from a United States port and tendering the bill of lading to the buyer, can be and frequently is performed by the seller tendering a bill of lading for soya beans that somebody else has shipped.'

So far as monies due under the contract is concerned, section 136 of the Law of Property Act 1925 reads in part:

'Any absolute assignment by writing under the hand of the assignor ... of any debt or other legal thing in action ... is effectual in law to pass and transfer ...:

(i)   the legal right to such a debt or thing in action
(ii)   all legal and other remedies for the same
(iii)   the power to give a good discharge for the same without the concurrence of the assignor ...'

For a legal assignment under this section, consent of the debtor is not required. All that is necessary is that:

(1)   It must be in writing
(2)   It must be signed by the assignor (i.e. the contractor)
(3)   It must be absolute and not by way of charge
(4)   The debtor i.e. the employer must be given notice of the assignment and the assignment is operative immediately on 'the date of the notice'.

Consideration between assignor and assignee is not necessary. Although the section refers solely to legal choses in action, the courts have held that it is equally applicable to equitable ones such as an interest in a trust fund.

Even where section 136 is not complied with, the courts will enforce equitable choses in action, subject to the requirements of section 53(1)(c) of the Law of Property Act 1925:

'A disposition of an equitable interest or trust subsisting at the time of the disposition must be in writing signed by the person disposing of the same.'

Notice to the debtor is not an essential to an equitable assignment and the assignment can be by way of charge. There can, therefore, be:

(1)   A legal assignment of a legal chose in action
(2)   A legal assignment of an equitable chose in action
(3)   An equitable assignment of a chose in action
(4)   An equitable assignment of an equitable chose in action.

However, in *Helstan Securities Ltd* v. *Hertfordshire County Council* (1978) the court was concerned with ICE Contract, 4th edition, which provided:

> 'Condition 3   The contractor shall not assign the contract nor any part thereof *or any benefit or interest therein or thereunder* without the written consent of the employer.' (author's emphasis)

It will be seen that the wording is considerably more emphatic and detailed than JCT 98 clause 19.1.1.

The contractors assigned monies due from the employer to Helstan Securities Ltd. Mr Justice Croom-Johnson held that, in view of the terms of the contract, there had been no valid assignment. He said:

> 'The clause is obviously there to let the employer retain control of who does the work. Condition 4, which deals with sub-letting, has the same object.
>
> Closely associated with the right to control who does the work is the right at the end of the day to balance claims for money due on the one hand to counterclaims. For example, for bad workmanship on the other.
>
> The plaintiffs say that such a counterclaim may be made against the assignees instead of against the assignors. But the debtor may only use it as a shield by way of set-off and cannot enforce it against the assignees if it is greater than the amount of the debt: *Young* v. *Kitchin* (1878).
>
> Why should they have to make it against people they may not want to make it against, in circumstances not of their choosing, when they have contracted that they shall not?'

That is a curious judgment. The debtor was the employer; the assignor was the contractor; the assignee was Helstan Securities.

Every assignee takes the assignment subject to the equities existing between the debtor and the assignor: so that the employer was entitled to withhold monies due to the contractor in respect of counterclaims or by way of set-off. The assignee can never be in a better position than the assignor.

Moreover, assignment of a chose in action, such as money due, in no way discharges the contractual obligations of the assignor. It is he and not the assignee, who remains liable on the obligations of the contract, so that if the employer's claim exceeds the amount of the debt, they could still sue the contractors.

Possibly, one contracting party can validly undertake to the other not to exercise his rights under the law of Property Act 1925 and in equity. But it is submitted that that can only be done expressly and it does not seem to have been achieved by the vague wording of JCT 98 clause 19.1.1.

3.10   *Indemnities given by the contractor*

The contractor provides indemnities to the employer against certain charges.

3.10.1   Fees and charges: clause 6.2

From this it would appear that any fees payable in respect of planning applications after the contract is entered into and those in respect of building inspection under the Building Regulations are to be paid by the contractor. But since the section also provides that such fees or charges are to be added to the contract sum, this appears to be no more than circuitous – unless they are caught by the exceptions which apply to:

(1)   Work or materials by local authority or statutory undertaker as nominated sub-contractor or supplier
(2)   Fees priced in the bills of quantities
(3)   Fees included as provisional sums in the bills of quantities.

In addition, and importantly, the contractor provides an indemnity to the employer against any costs which the employer may face as a result of the contractor failing to pay the fees or charges [1.19].

3.10.2   Royalties: clause 9.1

Any royalty charges payable in respect of processes, patents or inventions used in carrying out the Works are deemed to have been included in the contract sum and the contractor indemnifies the employer against any claims, proceedings and the like against the employer in connection with any infringement of patent rights. Clearly, and clause 9.2 sets this out, royalties payable as a result of architect's instructions must be added to the contract sum.

3.10.3   Injury to persons: clause 20.1

Not only is the contractor liable for injury or death arising out of the carrying out of the Works except in so far as caused by the negligence of the employer, he also indemnifies the employer against any claims, proceedings and the like **[1.19]**.

3.11   *Access to the Works and premises*

Clause 11 provides that the architect and his representatives must have access to the Works at all reasonable times. The reference to the architect's representatives is simply to ensure that the architect named in the contract in article 3 may nominate others to carry out the task of visiting the site. The architect named in the contract may be the name of a firm or the chief architect of an organisation. It is prudent, however, for the architect to formally nominate his representatives in writing to the contractor, so that they are not barred access. In addition, of course, it is common courtesy to do so. There is no express provision which entitles the contractor to object to such nomination, but a term to that effect would be implied for a compelling reason.

The clause also provides that access must be allowed to workshops and other places where work is being prepared for the contract. If any of the workshops or places belong to domestic or nominated sub-contractors (but strangely not suppliers), the contractor must include a term in the sub-contract to achieve a similar right of access. The clause uses the phrase 'so far as possible'. It is not clear why the contractor should not readily and successfully step down the requirement to sub-contractors. Indeed, it is already done in sub-contracts DOM/1 and NSC/C. The contractor has a further obligation to do 'all things reasonably necessary' to make the right effective. This is probably wide enough to require the contractor to institute proceedings against a sub-contractor if the architect or his representatives are refused admission to sub-contractor's premises.

The contractor or any sub-contractor may impose what clause 11 refers to as 'such reasonable restrictions' which are necessary to protect their proprietary rights. This refers to patent rights, design rights and trade secrets.

The draftsmen of JCT 98, as of JCT 80, apparently take the view that the contractor is not merely a licensee on the employer's premises but is in possession of them, because they create in clause 29 an obligation to permit the execution of work not forming part of the contract (clause 29.1); that is, if the contract bills have provided the information about such work which is necessary to enable the contractor to execute the Works in accordance with the contract. It is suggested that the information to be provided must be detailed enough so that the contractor knows precisely what is intended to be carried out by others and exactly how the work fits with work the contractor must carry out.

Where the contract bills do not provide that information, it appears that the contractor is entitled to refuse consent to other persons entering the site to execute work. If the employer requires some work to be done by others, the contractor may not unreasonably delay or withhold his consent (clause 29.2).

Statutory undertakers are discussed in **[4.14]**.

The issue is clouded considerably by clause 29.3 which provides that people so employed by the employer are deemed to be persons for whom the employer is responsible, not sub-contractors. Clause 20 gives the employer an indemnity against proceedings arising out of personal injuries or death except to the extent due to any act or neglect of the employer or anyone for whom the employer is responsible. Clause 29.3 refers to such persons as being 'deemed to be'. It is established that use of the word 'deemed' means that it is conceded that what is deemed is not actually the situation: *Re Cosslett (Contractors) Ltd, Clark, Administrator of Coslett (Contractors) Ltd (in Administration)* v. *Mid Glamorgan County Council* (1997). Therefore, the words in clause 29.3 mean that such a person is to be treated as one for whom the employer is responsible even if it is manifest to all that he is not. Therefore, even if such a person is a sub-contractor, he is to be treated, for the purposes of clause 20, as though he was not a sub-contractor.

It follows that the value of the contractor's indemnity under clause 20 could be considerably reduced if the employer is held responsible for those who are in fact sub-contractors to the main contractor but, because of clause 29.3, are deemed not to be so.

3.12   Antiquities

The contractor is obliged to report to the architect or the clerk of works all 'fossils, antiquities or other objects of interest or value' which he finds on the site including under the surface of the site: clause 34. He must use his best endeavours not to disturb them, including stopping work if necessary. In return, he is entitled to any loss and/or expense incurred by complying with this clause or any instruction issued by the architect. There is no provision for the contractor to apply for such loss and/or expense; it is the architect's task to take the initiative in considering the matter **[10.17]**.

3.13   Performance specified work: clause 42

This clause is, like many other clauses in JCT 98, perhaps needlessly long. Despite, or perhaps because of, its length, it contains several curious provisions and some omissions.

Under this contract, generally the contractor has no design liability. Indeed, in *John Mowlem & Co Ltd* v. *British Insulated Callenders Pension Trust Ltd* (1977), Judge Stabb said:

> 'I should require the clearest possible contractual condition before I should feel driven to find a contractor liable for a fault in the design, design being a matter which a structural engineer alone is qualified to carry out and for which he is paid to undertake, and over which the contractor has no control.'

The case was considering a piece of engineering design in respect of the watertight construction of a basement and the contract was JCT 63, but the principle holds good for JCT 98. In the more recent case of *Rotherham Metropolitan Borough Council* v. *Frank Haslam Milan & Co Ltd* (1996), the Court of Appeal was clear that it was the architect's job as expert and not that of the contractor to specify materials. The employer relied on the architect, not on the contractor.

If the employer wishes the contractor to carry out specific parts of the design, the JCT Contractor's Designed Portion Supplement should be incorporated. Its provisions are similar to a mini design and build contract such as the JCT Standard Form of Building Contract With Contractor's Design (WCD 98).

When should one use clause 42 rather than the Contractor's Designed Portion Supplement? Practice Note 25 deals with the topic. It suggests that elements such as trussed rafters, precast concrete floor units or simple installations for lighting, heating or power are suitable candidates, but that the system should not be used if it will substantially affect the final appearance or use of the building. The clause is obviously aimed at proprietary items such as the ones mentioned where there is design involved, but where it is not feasible for either the architect or a consultant to become involved. This is an attempt to address the problem where architects commonly specify such elements without giving too much thought to the situation which may arise if there is a design failure. Certainly in such circumstances, without special provision, the contractor is unlikely to bear any design responsibility despite the curious, and probably *obiter*, statement to the contrary in *Haulfryn Estate Co Ltd* v. *Leonard J. Multon & Partners and Frontwide Ltd* (1990) in relation to MW 80. In fact, there is probably no reason why clause 42 cannot be used to deal with rather more substantial pieces of design, if the elements are carefully chosen.

Performance specified work is defined, not in clause 1.3 with (most of) the other definitions, but in clause 42.1. There are four separate and indispensable requirements:

- The work must be identified in the appendix
- It must be for the contractor to carry out
- The requirements must be set out in the contract drawings
- The required performance must be stated in the bills of quantities which must also contain:
  either enough information to allow the contractor to price the work;
  or a provisional sum giving the position of the work in relation to the rest of the building and enough information so that the contractor can both programme the work and price the associated preliminaries (clause 42.7).

Unless the performance specified work is included in the bills of quantities in one form or another, it cannot be introduced by an architect's instruction after the contract documents are executed (clause 42.9). Moreover, clause 42.18 states that the work cannot be provided under a nominated sub-contract or a nominated supply contract. Clause 42.10 is similar in effect to clause 2.2.2.2 in that if there is an error or omission in the information which should be included in the bills of quantities referring to performance specified work, the error or omission is to be corrected to make it conform to the contract requirements and the correction is to be treated as a variation.

The contractor's standard is confined to exercising reasonable skill and care by clause 42.17. This is a lesser standard than would usually be required by the general law from a contractor carrying out design and build: *Viking Grain Storage Ltd* v. *T.H. White Installations Ltd* (1985). As if to emphasise this, the clause expressly, but in the circumstances needlessly, states that nothing in the contract is a guarantee of fitness for purpose. However, the contractor's liability for workmanship, materials and goods is undiminished. Moreover, the contractor is still responsible for the exercise of reasonable skill and care even if he sub-lets the design to others. That is only what the law would say in any event.

The clause sets out a detailed procedure. Clauses 42.2, 42.3 and 42.4 deal with the contractor's statement. This is a set of documents, which could be drawings and/or specifications, which the contractor must prepare and submit to the architect before carrying out any performance specified work. On a strict reading of this clause, the contractor who has, say, three items of performance specified work in the contract must provide statements for all of them before commencing work on any one of them, notwithstanding that the architect may not have issued any instructions as to the expenditure of a provisional sum included for one of the items. This is a trap which architects will have to be vigilant to avoid. The statement must have been seen by the planning supervisor and must deal with any unfavourable comments and it must be detailed enough so that the architect can understand what is proposed. If the architect wants the statement to contain specific information, a note to that effect should be included in the bill of quantities description or in the architect's instruction.

The architect, under clause 42.13, can request the contractor to provide an analysis of the performance specified work if the bills of quantities do not show the analysis. The contractor has 14 days to comply.

A very important part of clause 42.3 requires the contractor to provide the statement in reasonable time to enable the architect to produce any information which he is to provide under clause 5.4. This is to deal with any interface between architect and contractor design. Clause 42.14 requires the architect to give any necessary instructions for the integration of the performance specified work a reasonable time before it is due to be executed. Obviously, this is another factor which the contractor must bear in mind in deciding what is a reasonable time in which to supply the statement. A failure by the contractor to provide the statement in time would amount to a breach of contract for which the employer would be able to claim any resultant damages. It is difficult to see what damages would be suffered by the employer, because the likely result of the contractor's delay in providing the statement would be to delay the Works. The result of that would be that the employer would be able to recover liquidated damages for the period and the contractor would be unable to claim anything, being the author of his own misfortune.

If interfacing with the architect's design is not a consideration, the contractor's obligation is to provide the statement either by the date stated in the bills of quantities or in the architect's instruction, or a reasonable time before the contractor is to carry out the work (clause 42.4). Although a reasonable time will depend upon the circumstances, one of the circumstances will be the architect's right to comment under clause 42.5 after which the contractor must provide an amended statement. Curiously, although clause 42.16 expressly prevents the contractor from being given an extension of time if the contractor has not complied with his obligations regarding time set out in clauses 42.4 and 42.5, no such express prohibition is placed in regard to the contractor's failure to provide the statement within a reasonable time as stated under clause 42.3.

The architect has 14 days in which to give written notice to the contractor if in the architect's opinion the statement does not properly explain the proposal. If the architect does give the notice, the contractor must amend the statement and re-submit. The last part of clause 42.5 makes clear that whether or not the architect asks for clarification will not affect the contractor's responsibility for the statement or, more importantly, for the work. This warning is repeated at the end of clause 42.6 which places an obligation on the architect to give notice to the contractor if he finds what he believes to be a 'deficiency' (as the clause quaintly calls it) in the contractor's statement. The notice ought to be in writing although the clause does not expressly require it.

Although clause 42.6 is clear that the architect is obliged to give notice of any deficiency in the statement, the contract is silent about the next step. In short, there is no requirement that the contractor must take any action as a result. Neither does the contract say what the architect can do if the contractor ignores the notice. It is little consolation to the employer simply to be told that the contractor is responsible for the problem if the work does not in fact satisfy the specification. The architect can issue an instruction under clause 42.11 requiring a variation to change the performance specified work proposed by the contractor, provided that does not have the effect of increasing the work without the agreement of both employer and contractor (clause 42.12). There is no express provision for the architect to deal with a failure by the contractor to satisfy the performance specification. Once constructed, it is clearly work which is not in accordance with the contract and it can be dealt with under clause 8.4, but it is nonsensical to wait until that point. It is easy to say that it is highly unlikely that any conscientious contractor would ignore the architect's notice so as to precipitate such a situation, but it is exactly such highly unlikely events which seem to occur with enough frequency to keep adjudicators, arbitrators and judges busy.

Clause 2.4.2 makes clear that if the architect finds a discrepancy in the contractor's statement, the contractor must correct it at no additional cost to the employer. Although the contractor must notify the architect about the correction, it is noteworthy that he is not obliged to carry out the correction to the architect's satisfaction.

Although the architect is entitled to give instructions under clause 4 [2.9] and instructions requiring a variation under clause 13 [6.2], the contractor need not immediately comply with the instruction if he believes that to carry out the instruction would adversely affect the 'efficacy' of the performance specified work. Provided he acts within seven days of receiving the instruction, he may set out the adverse effect in writing to the architect. The result is that the architect may either act to remove the problem from his instruction or the instruction is of no effect without the consent of the contractor. Although the withholding or delay of consent is qualified by reference to it not being unreasonable, that is presumably the very reason why the contractor gave notice in the first place. Effectively, therefore, it appears that the contractor can block any architect's instruction which affects performance specified work.

3.14 Does the contractor have a duty to warn?

Without express provision to the contrary, for example by use of the Design Portion Supplement or clause 42 performance specified work, the contractor has no liability for design. Clause 2.1 makes the position clear that his obligation is simply to carry out and complete the Works in accordance with the contract documents. Does that mean that he is entitled to blindly build what the drawings and specifications set out, even if there are obvious errors? Most architects and sound common sense would say not, but the substantial case law on the topic seems to be inconsistent.

Some years ago, a Canadian case established that a contractor will be liable to the employer for building errors in a design if the original architect was not involved in the construction stage: *Brunswick Construction* v. *Nolan* (1974).

In two cases in 1984 (*Equitable Debenture Assets Corporation Ltd* v. *William Moss* and *Victoria University of Manchester* v. *Wilson and Womersley*) a court came to the conclusion that contractors did have a duty to warn the architect if they believed that there was a serious defect in the design. Subsequently, however, in *University of Glasgow* v. *William Whitfield & John Laing (Construction)* (1988) another court decided that if the contractor did have a duty, it was to the employer and only in those exceptional cases where the contractor knew that the employer was relying on him for at least part of the design. But then, in *Lindenberg* v. *Canning* (1992), it was concluded that a contractor had a duty to at least raise doubts with the architect if there appeared to be something wrong with the drawings.

The Court of Appeal case, *Plant Construction plc* v. *Clive Adams Associates and JMH Construction Services Ltd* (2000), took the position one stage further. Although concerned with sub-contract work, the principles are equally applicable to main contracts. JMH was involved in the design of temporary support work to a roof. Their design was overruled by the employer's engineer. JMH warned the engineer of the danger of his design, but to no avail. The roof collapsed and the court held, not only that JMH had a duty to warn, but that it had not warned with sufficient force. It is difficult not to feel a great deal of sympathy with JMH in this case.

Probably the contractor's duty to warn only arises if the design is seriously defective. In the *Plant Construction* case, it seems to have been a potential danger to life. A contractor who did not warn an architect who had made an error in an eaves detail would be unlikely to have any liability. The test seems to be whether the employer in fact relies upon the contractor and whether he is entitled to do so. Where it can be shown that the employer does rely, even partly, on the contractor, there will be a duty to warn of serious defects. Cases where the duty arises to warn the architect will be rare, because the architect seldom, if ever, relies or should rely on the contractor. In the context of JCT 98, the duty is extremely limited, because the employer will usually be relying on the architect and not the contractor. The contractor is certainly not charged with checking the architect's work.

CHAPTER FOUR

SUB-CONTRACTORS, SUPPLIERS AND STATUTORY UNDERTAKERS

4.1 Assignment and sub-letting

Sub-contracting and assignment are often confused. Before considering the contract provisions in detail, it is important to understand the difference between these terms. They were set out with admirable clarity by Lord Justice Staughton in *St Martins Property Corporation Ltd and St Martins Property Investments Ltd* v. *Sir Robert McAlpine & Sons Ltd and Linden Gardens Trust Ltd* v. *Lenesta Sludge Disposals Ltd, McLaughlin & Harvey PLC, and Ashwell Construction Company Ltd* (1992) in the Court of Appeal **[3.9]**.

4.2 Assignment

Clause 19.1.1 restricts the assignment of the contract by either party without the written consent of the other. In the *St Martins* case, the House of Lords (1993) held that this clause is effective in preventing the benefit of the contract being assigned. The employer might wish to sell the building before the final certificate is issued or the contractor may wish to assign the right to receive payment in return for a cash advance from a funder. Clause 19.1.1 provides that consent has to be given by the other party in each case. A party may probably refuse consent on grounds which might be considered unreasonable. This can pose real problems and if the employer may wish to assign the benefit of the contract (i.e. sell or otherwise transfer the property to another) before practical completion, an amendment to the clause is advisable.

Assignment of the burden of a contract is not allowed under the general law, so this clause is superfluous in that regard. For the contractor to effectively transfer to another the duty to carry out the work set out in the contract documents, or for the employer to transfer the duty to pay for such work, the consent of both parties to the contract would be needed together with the consent of the party who is to shoulder the burden in place of either contractor or employer.

Clause 19.1.2 deals with the situation where the employer sells the freehold or the leasehold interest in the premises comprising the Works to a third party, or where he grants a leasehold interest in the premises. It will only apply if so stated in the appendix. In any of these instances, the employer may assign to that third party the right to bring proceedings in the employer's name and to enforce any of the terms of the contract. This clause does not give the employer the right to sell the premises before he has received them from the contractor at practical completion, therefore it does not conflict with clause 19.1.1. However, once the employer has received the building and disposed of it by sale or lease, it enables the purchaser to act as if he was the employer so far as the benefits of the contract are concerned. For example, the obligation to pay the contractor remains with the employer, but the purchaser can enforce the defects liability provisions. There is a proviso that the third party cannot dispute any agreement which is legally enforceable and which is entered into between the employer and the contractor before the assignment. If the employer and the contractor have entered into an agreement under which the contractor is not obliged to make good certain defects and no monetary deduction is to be made, it is binding on the purchaser of the premises under clause 19.1.2.

4.3   Sub-letting

Clause 19.2.1 defines a 'domestic sub-contractor' as a person, other than a nominated sub-contractor, to whom the contractor sub-lets any part of the Works. Clause 19.2.2 provides that the contractor must not sub-let without the architect's consent. In contrast to assignment, the architect may not unreasonably delay or withhold consent. There is no requirement that the contractor must inform the architect of the names of sub-contractors. It is merely consent to the fact of sub-contracting which is required. It may be reasonable for the architect to refuse to give consent until the name and perhaps other details of the prospective sub-contractor are made known. If the architect does give his consent, the contractor's obligations under clause 2.1 are not affected. In practice, the architect will normally give his consent if the contractor satisfies him that the proposed sub-contractor is capable of doing the work in accordance with the contract and the contractor's programme.

If the contractor has become insolvent when a defect is found in a sub-contractor's work, the employer has no contractual remedy and, following *Murphy* v. *Brentwood* (1990), little hope of a tortious remedy. This may pose difficulties at that late stage. To overcome such problems, the architect may make the provision of an acceptable warranty on the part of all sub-contractors a pre-condition to the giving of any consent to sub-letting. A better solution is for the architect to insert such a stipulation, accompanied by an example of the warranty required, in the specification or bills of quantities.

4.4 Listed sub-contractors

Clause 19.3 provides a useful alternative to the use of nominated sub-contractors **[4.6]**. It is a means whereby the employer can limit the prospective sub-contractors to a chosen few. That is the theory although, in practice, that may not be the result. The resultant sub-contractor is a domestic sub-contractor (clause 19.3.3). Clauses 19.3.1 and 19.3.2 set out the criteria:

- The bills of quantities must specify certain work as required to be carried out by persons selected by the contractor from a list in or attached to the bills of quantities.
- There must be no less than three persons on the list which can be augmented by names added by the employer and/or the contractor with the consent of the other at any time until the sub-contract is executed (clause 19.3.2.2). Consent may not be unreasonably delayed or withheld and this effectively means that it will be difficult for the employer to block any names submitted by the contractor unless they are demonstrably unsuitable. To avoid the list dropping below the minimum, it is good practice for the initial list to contain five names.
- If the number on the list falls below three, the employer and the contractor must agree the addition of further names or the contractor may carry out the work. In doing so, the contractor may, if he wishes, sub-let to any sub-contractor in accordance with clause 19.2.

Once the contractor has entered into a sub-contract with someone on the list, the architect has no further involvement in extensions of time, financial claims or determination of the sub-contractor's employment.

4.5   *Sub-contract provisions*

Clause 19.4 sets out certain provisions as a condition to sub-letting. Clause 19.4.1 states that each sub-contract must provide that the employment of the sub-contractor determines immediately determination of the contractor's employment takes place. Clause 19.4.2 states that the sub-contract must include certain provisions regarding unfixed materials and goods delivered to the Works or adjacent to the Works. There are four such provisions:

- Such materials and goods must not be removed without the contractor's consent unless for use on the Works
- Ownership of such materials and goods is to be automatically transferred to the employer after the value has been included in an interim certificate for which payment has been made
- If the contractor has paid for such materials and goods before certification and being himself paid by the employer, ownership passes to the contractor
- The operation of this clause is not to affect ownership in off-site materials passing to the contractor as provided in clause 30.3 of JCT 98.

The architect should try to ensure that these provisions are included in sub-contracts and he is perhaps entitled to refuse to consent to sub-contracting unless evidence of such inclusion is produced. However, the contract does not place any express duty on him to check. Although standard sub-contract DOM/1 contains such provisions, many contractors habitually sub-contract using their own terms which not only do not contain such provisions, but also do not create a satisfactory 'back to back' sub-contractual arrangement. Even if such provisions are included, they are ineffective to safeguard the employer from the perils of retention of title, if the sub-contractor has bought the goods himself on terms that the supplier retains ownership until payment is made. Building contract chains are so long that it is virtually impossible to check down to the ultimate supplier that ownership has passed unimpeded up to the contractor. Breach of the provisions of clauses 19.1.1 or 19.2.2 is a ground for determination by the employer under clause 27.2.1.4 although it may be a draconian remedy in most circumstances **[9.4]**.

4.6 *Nominated sub-contractors*

Nomination is a case of the employer trying to have everything his own way. He wants to stipulate who the contractor must employ to carry out certain work, but at the same time, he wants no responsibility. Nomination often causes more problems for the employer than for the contractor. It is something best avoided if possible. Sometimes, however, it is necessary that a specialist sub-contractor is nominated for particular work. On those occasions, it is essential to operate the contractual mechanisms correctly and precisely. Anecdotal evidence suggests that it is rare for this to happen.

Clause 35 is long and complex in words and in operation. It cannot be stressed too much that the nomination procedures laid down must be precisely followed. Even when that is done and all the appropriate pieces of paper are in place, the procedures for practical completion, extension of time, payment and renomination are real challenges to comprehension.

Clause 35.1 provides that nomination can take place in any of eight ways. The architect may either use a prime cost sum or name the sub-contractor:

- In the bills of quantities
- In an instruction regarding the expenditure of a provisional sum
- In an instruction requiring a variation under clause 13.2
- By agreement with the contractor.

Clause 35.1 is very important. It states that if the architect has reserved the choice of sub-contractor to himself in this way, the sub-contractor 'shall be nominated in accordance with the provisions of clause 35 and a sub-contractor so nominated shall be a Nominated Sub-Contractor for all the purposes' of the contract. A question which frequently arises is whether a sub-contractor whom the architect invites to tender in an informal way becomes a nominated sub-contractor for all the purposes of the contract if the architect subsequently instructs the contractor to accept the tender and enter into a sub-contract. The extract quoted is highly ambiguous when read in that light. It has two possible meanings:

- The sub-contractor must go through the nomination procedure in clause 35, following which he is referred to as a nominated sub-contractor for all the purposes of the contract;
  or

- The sub-contractor chosen by the architect will be considered nominated under clause 35 and will be referred to as a nominated sub-contractor for all the purposes of the contract.

Although the courts have shown themselves willing to consider, as nominated, a sub-contractor who has been treated as nominated by both architect and contractor, albeit the sub-contractor has not gone through the clause 35 procedure (*St Modwen Developments Ltd* v. *Bowmer & Kirkland Ltd* (1996)), it is thought that the first interpretation is probably the correct one, in view of clause 35.3 which stipulates that nomination to which clause 35.1 applies must be carried out in accordance with clauses 35.4 to 35.9. Architects must beware of trying to avoid the complexities of nomination, while achieving the effect of nomination by other means. They may find themselves in unchartered waters.

The contractor is protected from the imposition of an unsuitable sub-contractor by clause 35.5 which entitles him to make a reasonable objection in writing within seven working days of receipt of the architect's nomination instruction under clause 35.6. If the objection is reasonable, the architect may do one of the following:

- Issue instructions removing the grounds for objection
- Cancel the nomination and
  either
  omit the work under clause 13.2
  or
  nominate another sub-contractor.

Clauses 35.20, 35.21 and 35.22 should be read together. They spell out certain limits to the relationship between the nominated sub-contractor and the employer and the contractor.

Clause 35 does not make the employer liable to the nominated sub-contractor in any way other than through the NSC/W warranty agreement. Moreover, the extent to which the liability of the sub-contractor may be limited to the contractor under NSC/C clause 1.7 will also limit the contractor's liability to the employer. This potentially restricts liability in the circumstances specified in NSC/C clause 1.7 right up the contractual chain from suppliers or sub-sub-contractors.

Clause 35.21 bears careful reading. The contractor is not liable for any element of design, whether it be actual design, or selection of materials or the satisfaction of a performance specification, to the extent that the nominated sub-contractor has already designed, selected or satisfied. Moreover, the contractor is not liable to the employer for providing the information which NSC/W requires from the nominated sub-contractor in order to allow the architect to comply with clauses 5.4.1 and 5.4.2.

4.7   *Contractor doing nominated sub-contract work*

If the architect agrees and the type of work is that which the contractor normally carries out, the contractor may be allowed to submit a tender (clause 35.2.1). The work must be identified in the appendix. The employer is not obliged to accept the contractor's tender for such work, but if it is accepted, the contractor cannot sublet the work without the architect's consent. In this instance, the withholding of consent is not qualified so as to be reasonable. Clauses 35.2.2 and 35.2.3 stipulate that clause 13 applies to the work in the tender and none of the remainder of clause 35 will apply. That is just common sense.

A serious bar to the operation of this provision in practice is that, because of the lengthy nomination procedures, it makes good sense for the architect to invite tenders for most of the nominated subcontract work before entering into the main contract.

4.8   *The procedure*

The relevant forms are: NSC/T Tender (parts 1, 2 and 3), NSC/W Warranty, NSC/N Nomination, NSC/A Articles and NSC/C Conditions.

The procedure is contained in clauses 35.6 and 35.7 and is complex, but briefly it will amount to the following:

- The architect will have sent the completed NSC/T, Part 1 (the invitation to tender) to the prospective sub-contractor together with a blank Part 2 (the tender document), the drawings and/or specification and/or bills of quantities (the 'numbered tender documents') and the appendix to the main contract as it is envisaged to be completed. In addition, the architect must include a copy of NSC/W (the collateral warranty) with the contract details completed.
- The sub-contractor must complete NSC/W and Part 2, sign each and return them to the architect.

- If the employer approves, he must sign Part 2 and enter into NSC/W agreement.
- The architect then sends to the contractor a nomination instruction NSC/N, NSC/T Part 1 and 2, copies of the numbered tender documents, the principal contractor's health and safety plan, NSC/W, any changes to items 7, 8 and 9 of NSC/T Part 1 and any other documents or amendments which 'have been approved by the Architect'. The other documents or amendments clearly refers to any qualifications or changes which the sub-contractor has proposed and which have been accepted by the architect. At the same time, the architect must send to the sub-contractor a copy of NSC/N and the main contract appendix as actually completed.
- The contractor and sub-contractor are to agree the particular conditions (Part 3) and to enter into a sub-contract on NSC/A (the articles of agreement which incorporate conditions NSC/C) within 10 working days of receipt of the nomination instruction and to send the architect a copy of the completed NSC/A. Not for the first time, the contract refers to 'working days' and, although the meaning is clear, it seems an unnecessary complication in view of the fact that periods of time have already been defined quite precisely in clause 1.8.

Clauses 35.8 and 35.9 provide that if the contractor, having used his best endeavours to agree Part 3 and to enter into NSC/A, fails to do so, he must inform the architect either:

- *The date he expects to complete NSC/A* and the architect may consult the contractor and fix whatever new date he considers reasonable.
- *That the failure is due to other matters which must be specified* and the architect may:
  - issue further instructions to allow completion
  - cancel the nomination and omit the work
  - cancel the nomination and nominate another sub-contractor
  - if he does not consider the matters justify the failure to complete NSC/A, notify the contractor accordingly who must then agree the particular conditions and enter into the sub-contract.

It is not clear how the architect is to operate the last option, because if the contract cannot impose an obligation on two parties to agree, much less can the agreement be imposed by an architect's instruction. In any event, the sub-contractor is expressly entitled to withdraw his tender under NSC/T, Part 2, either within seven days of notification of the identity of the main contractor or if he is 'for good reasons' unable to agree Part 3 Particular Conditions. The general law would entitle him to withdraw his tender at any time before acceptance whatever he may have said about leaving it open for any particular period unless payment or some other consideration was given by the employer.

If the architect cannot, as a matter of practice and law, operate the last option, it seems he will be obliged to alter his position by issuing further instructions, omission of work or renomination if the contractor and the sub-contractor cannot agree. It is difficult to see why the architect should wish to omit the work altogether. If it is important enough to warrant nomination, presumably it will have to be carried out in some way.

4.9   *Payment provisions*

The payment provisions are to be found in clauses 35.13, 35.17, 35.18 and 35.19. Clause 35.13.1 provides that the architect must direct the contractor regarding amounts due to nominated sub-contractors which are included in interim certificates and inform the sub-contractors to whom payment is to be made. The contractor is obliged to discharge each payment as set out in the nominated sub-contract (NSC/C) (clause 35.13.2).

The only circumstances in which the employer may pay any sub-contractor directly are set out in clauses 35.13.3, 35.13.4 and 35.13.5. Following the issue of any certificate containing amounts due to a nominated sub-contractor, the contractor must provide the architect with reasonable proof of payment of the nominated sub-contractor before the issue of the next certificate. This provision is much misunderstood and merits careful reading.

If the contractor fails to provide the reasonable proof, the architect has no choice but to certify the fact, presumably to the employer with a copy to the contractor. Clause 35.13.5.1 expressly states that a copy must be sent to the relevant nominated sub-contractor. This certificate is a condition precedent to the employer's duty to pay the sub-contractor directly. Once the certificate has been issued, however, the employer is, in principle, obliged to pay the nominated sub-contractor the amount withheld by the contractor and to deduct the amount from the sum due to the contractor on the next certificate. This statement of principle is subject to important provisos.

Clause 35.13.5.2 makes clear that the deduction from the amount due to the contractor is to be made *after* any other deductions which the employer is entitled to make; for example, liquidated damages. The employer is not obliged to pay to any nominated sub-contractor a sum greater than the amount which would have been paid to the contractor after such deductions. Therefore if a nominated sub-contractor is owed £20,000 under a certificate but is not paid by the contractor, and if the next certificate is worth a total of £25,000, but subject to a deduction of £10,000 by the employer, the nominated sub-contractor will only receive £15,000. From this and the period allowed to the contractor to provide the reasonable proof, it is obvious that a nominated sub-contractor who is not paid by the contractor will not be paid by the employer under these provisions until the date the contractor should have been paid the amount in the certificate following the one containing the original amount. This is confirmed by clause 35.13.5.3.1. Unusually, both the amount of the deduction and the direct payment to the nominated sub-contractor are said to be inclusive of VAT. There are other provisos:

- If the amount due to the contractor under a certificate is retention, the amount deductible and, therefore, available for payment to the sub-contractor must not exceed the retention due to the contractor. Retention sums in respect of other nominated sub-contractors are untouchable in this connection.
- If more than one nominated sub-contractor is to be paid by the employer, but insufficient money is available to be deducted from the next certificate, the employer may either proportion the payments to the respective amounts owing or adopt some other method which, in all the circumstances, the employer believes is fair and reasonable.
- The provisions for direct payment cease to apply if, 'at the date when' the deduction from the payment to the contractor and payment to the sub-contractors would have been made, a petition has been presented to the court for winding up the contractor or a resolution has been passed for his winding up, except for amal-gamation or reconstruction.

Clause 35.13.4 explains the position if the contractor cannot provide reasonable proof due to the sub-contractor's failure to provide the evidence needed by the contractor. Such evidence may be in the form of a receipt signed by the sub-contractor. If the architect is satisfied that the sub-contractor's failure is the only reason for the contractor's inability to provide the necessary proof, the situation is treated as if the contractor had provided the proof. It is not easy to see how the contractor is to convince the architect in the absence of a signed receipt – perhaps by showing the architect a cheque stub or a copy of the cheque itself from the bank.

What if the contractor perfectly properly exercises his right to set off from the nominated sub-contractor under NSC/C clause 4.16.1.2? To the extent that he has set off, he clearly has not paid the amount due to the nominated sub-contractor in the interim certificate. The answer is contained in clause 35.13.2 which, as noted above, repays careful reading. It requires the contractor to pay 'in accordance with Conditions NSC/C'. Since NSC/C contains the set-off provision, a payment is made in accordance with NSC/C if it is made subject to set-off properly made and if the requisite notices have been issued. How far the architect should become involved in such set-off is difficult to say, but he is clearly required to satisfy himself that the contractor has complied with his obligations. Part of the 'reasonable proof' will be evidence to show why the full amount has not in fact been paid.

Clause 30.7 stipulates that not less than 28 days before the issue of the final certificate, the architect must issue an interim certificate which must include the final account sums for all nominated sub-contractors. However, clause 5 of NSC/W gives the nominated sub-contractor a valuable right to early final payment provided it has not been amended. The architect has until 12 months from the date of practical completion in which to issue an interim certificate containing the final amount (clause 35.17). This is subject to two provisos:

- The architect must be satisfied that the nominated sub-contractor has remedied any defects which he was obliged to remedy; *and*
- The nominated sub-contractor must have sent to the architect or the quantity surveyor all the documents necessary for the final sub-contract account.

Early final payment does not relieve the contractor of his responsibilities. Until practical completion of the Works or the date when the employer takes possession of the Works, if earlier, the contractor is responsible for loss or damage to the sub-contract works, for which early final payment has been made, to the same extent that he is responsible for the rest of the Works (clause 35.19). In addition, the Works' insurance clauses remain in full effect.

The question of defects has, quite rightly, exercised the minds of the JCT draftsmen and the results of their deliberations are found in clause 35.18 which considers the situation if, before the issue of the final certificate, the sub-contractor does not remedy any defect for which he is responsible. In these circumstances, the contract introduces the concept of the 'substituted sub-contractor' to be nominated by the architect for the sole purpose of dealing with the defects. In an almost throwaway provision, clause 35.18.1.1 states that 'all the provisions' of clause 35 will apply to this nomination. It is reasonable to assume that the words mean precisely what they say, but it is not clear whether that means that the full procedure for nomination must be followed together with all the provisions for practical completion and extensions of time. That would be a most unwieldy procedure and in most instances a quite unnecessary complication.

Whatever the position with regard to that, it is important that the contractor agrees the substituted sub-contractor's price for doing the work. Indeed, the contractor is not entitled to delay or withhold his agreement unreasonably. The reason for the importance is that, although the employer has a duty to attempt to recover the cost of rectification from the original sub-contractor under the provisions of NSC/W, if he fails or partly fails, the contractor must pay the employer what he fails to recover.

Clause 35.13.6 deals with the situation where, under NSC/W, the employer has paid money to the nominated sub-contractor, before nomination, in order to make an early start on work which is part of the nominated sub-contract sum or tender sum. This is likely to be, but not necessarily, design work or the ordering of materials. The employer (not the architect) is to send the contractor the nominated sub-contractor's written statement of the amount to be credited to the contractor and the employer may then withhold up to that amount from amounts stated as due in interim certificates. The appropriate withholding notices must be sent **[5.21]**. The amounts withheld may not, for obvious reasons, exceed the amount due to the relevant sub-contractor in any one certificate.

4.10   *Practical completion, extension of time and failure to complete*

Clause 35.16 provides that the architect must certify when practical completion of the nominated sub-contract works has taken place and the sub-contractor has complied sufficiently with his obligations under clause 5E.5 of NSC/C (CDM Regulations). The clause curiously proceeds to state that practical completion of such works is then 'deemed' to have taken place on the day named. The implications of this are discussed in **[12.1]**.

If the employer has taken partial possession under clause 18 of a part of the Works which includes some nominated sub-contract work, practical completion of such nominated sub-contract work is deemed to have occurred on the relevant date. The written statement which the architect is to issue under clause 18.1 must be copied to any nominated sub-contractor concerned.

Clause 35.14 is generally misread or at any rate misunderstood by the construction industry, architects, quantity surveyors and contractors included. Yet the clause is relatively brief and straightforward. Under NSC/C, clause 2.2.1, the sub-contractor is entitled to an extension of time on much the same grounds as the contractor under the main contract, but with the addition of the contractor's acts, omissions or defaults. Despite what some commentators seem to think, clause 35.14 does not require the architect to decide the length of extension of time due to a nominated sub-contractor except in the most indirect and broad way.

The process is that the sub-contractor must give written notice to the contractor of delays together with relevant details, similar to the details required of the contractor under clause 25 of JCT 98. NSC/C clause 2.2.3 states that the contractor must join with the sub-contractor in asking for the architect's consent if the sub-contractor so requests. The significance of this lies in the sub-contractor's power to require the contractor to allow his name to be borrowed to enable an arbitration to take place, effectively between the nominated sub-contractor and the employer. The contractor, however, must obtain the architect's consent before granting an extension. Clause 35.14.1 requires the contractor to follow NSC/C provisions requiring the architect's consent in granting extensions of time. Clause 35.14.2 imposes a duty on the architect to operate the appropriate NSC/C clauses when he receives the details from the contractor and a joint request for his consent. 'Consent' has an entirely different meaning to 'decide'. It means to 'allow' or 'permit' or 'concur'. The architect is not called upon to decide the extension of time, but clearly to consent or otherwise to an extension of time proposed by the contractor. It is obvious that, in conveying details of the nominated sub-contractor's delay to the architect, the contractor must propose an extension of time even though neither JCT 98 nor NSC/C refers to it. He cannot consent, permit, allow or concur in a vacuum, i.e. he must have something to which he can consent.

That is clear from the plain meaning of the words used. It also makes good common sense that while the architect might be useful to curb cases of unreasonable decisions by the contractor, he is certainly not in a position to be able to make precise determinations of the period of extension warranted by circumstances of which only the contractor and the nominated sub-contractor have a detailed knowledge. However, the architect will be able to take a broad view of the position on the basis of his own knowledge of the project and the details submitted. If he decides that the contractor is proposing too little extension or, perhaps more unusually, too much, he simple withholds his consent. This is a powerful incentive for the contractor to get his estimate right, because if he does not get the architect's consent, it follows that the architect cannot form a view that the extension of time procedures have been properly operated. It is important that the architect does form that view if the contractor is hoping to recover damages from the sub-contractor for late completion. NSC/C has a similar review process to JCT 98, the period expiring 12 weeks after practical completion of the sub-contract works as certified by the architect under clause 35.16. The architect's consent is also required for the review process.

If the sub-contractor fails to complete his work during the sub-contract or extended period, clause 35.15 provides that the contractor may again notify the architect who, *if he is satisfied that the extension of time procedures have been properly applied,* must issue a certificate of non-completion not later than two months after the date of notification. The issue of this certificate enables the contractor to recover damages from the sub-contractor which will, in appropriate instances, include but not be limited to liquidated damages which the contractor is liable to pay to the employer under the main contract.

4.11   Renomination

Renomination is covered by clauses 35.24, 35.25 and 35.26. Clause 19.5.2 makes clear what was uncertain in standard contracts before *North West Metropolitan Regional Hospital Board* v. *Bickerton* (1970); namely that the contractor is not required to supply and fix materials nor to execute work which is to be carried out by a nominated sub-contractor. Of course the parties may agree otherwise and clause 35.2.1 makes particular reference to the contractor's entitlement to submit a tender provided that certain criteria are met [4.7]. A number of important principles have been established by the courts:

- *Percy Bilton* v. *Greater London Council* (1982):
  - Loss arising from failure or withdrawal of the nominated sub-contractor must be borne by the contractor.
  - The employer must renominate within a reasonable time of notification from the contractor. Failure to do this entitles the contractor to extension of time and gives him grounds for making a claim against the employer for any loss caused by unnecessary delay to renomination (clause 35.24.10).
- *Fairclough Building Ltd* v. *Ruddlan Borough Council* (1985)
  - A renomination instruction must include both remedial and uncompleted work.
  - A contractor can refuse to accept renomination of a substituted sub-contractor who does not offer to complete his part of the work within the overall period for the contract.

Renomination will become necessary if:

- The contractor is of the opinion that the nominated sub-contractor has defaulted in certain specified matters and the architect agrees (clause 35.24.1)
- The sub-contractor becomes insolvent (clause 35.24.2)
- The sub-contractor determines his own employment due to certain specified defaults of the contractor (clause 35.24.3)
- The sub-contractor's employment has been determined for corruption (clause 35.24.4)
- In the case of re-execution of work which the architect is empowered to order under clauses 7, 8.4, 17.2 or 17.3, but which the nominated sub-contractor cannot be required and does not agree to carry out (clause 35.24.5).

For the sake of brevity, these will be referred to as: sub-contractor's default, insolvency, contractor's default, corruption and sub-contractor's option, respectively. They are discussed in detail below.

4.11.1   Sub-contractor's default (NSC/C clauses 7.1.1.1 to 7.1.1.4)

This is a complicated procedure. It is triggered if the contractor notifies the architect that the sub-contractor has defaulted for any of the following reasons:

- Completely suspending the sub-contract work without reasonable cause;

- Failing to proceed regularly and diligently with the work
- Refusing or neglecting to comply with the contractor's written notice to remove defective work or materials and as a result the works are substantially affected
- Failing to comply with the sub-contract provisions with regard to assignment and sub-letting.

The contractor must include the sub-contractor's observations. If the architect is reasonably of the opinion that the sub-contractor has defaulted, he must instruct the contractor to serve notice on the sub-contractor specifying the default (clause 35.24.6). The architect may state that the contractor must obtain a further instruction before determining the sub-contractor's employment. The contractor must notify the architect when determination has taken place.

The architect must renominate within a reasonable time. He is entitled to invite tenders for the work in order to achieve a fair price. The renomination must include the doing or redoing of work and the supply or resupply of materials as necessary.

If the grounds for determination are failure to comply with the contractor's written notice to remove defective work or materials or failure to correct defects, the contractor is entitled to agree the price of the substituted sub-contractor. The contractor, however, cannot unreasonably withhold his agreement. This provision is important because if the contractor is unable to recover from the original sub-contractor the extra cost of the substituted sub-contractor, the contractor must pay the difference **[4.7]**.

4.11.2   Insolvency (NSC/C clause 7.2)

In the case of what might be collectively termed insolvency, determination is automatic or as a requirement of the employer respectively. Legal advice must always be sought in the case of determination due to insolvency because the position is very complicated. Clause 35.24.7 provides that if the receiver, administrative receiver or administrator is willing to carry on the sub-contract, the architect may postpone the renomination if it will not prejudice the interests of either the employer, the contractor or any other sub-contractor.

4.11.3   Contractor's default (NSC/C clause 7.7)

Clauses 35.24.8.1 provides that renomination is triggered if the nominated   sub-contractor   determines   his   own   employment because the contractor has defaulted by completely suspending the main contract Works without reasonable cause or by failing to proceed with the main contract Works regularly and diligently and as a result reasonable progress of the sub-contract works is seriously affected.

The architect must renominate within a reasonable time including the doing or redoing of work and the supply or re-supply of materials. When the amount due to the substituted sub-contractor has been certified, the employer may deduct the extra he has had to pay from any money due to the contractor or he may recover it as a debt.

4.11.4   Corruption (NSC/C clause 7.3)

The architect must renominate within a reasonable time including the doing or redoing of work and the supply or re-supply of materials if the nominated sub-contractor's employment is determined due to corruption (clause 35.24.7.4).

4.11.5   Sub-contractor's option (NSC/C clause 3.8)

In this instance, clause 35.24.8.2 stipulates that the architect must make the necessary renomination to carry out the work to be re-executed.

4.11.6   Payment

Payment of the nominated sub-contractor is dealt with in clauses 35.13, 35.17, 35.18 and 35.19 [4.9]. The aftermath of renomination is covered in clause 35.24.9.

Amounts payable to the substituted sub-contractor must be included in interim certificates and added to the contract sum. The architect must also direct the contractor regarding payment. As far as the original sub-contractor is concerned, he must pay or allow the contractor direct loss and/or damage caused by the determination. The contractor need make no further payments to him until after completion of the sub-contract work. At that stage, the original sub-contractor may apply for payment to the contractor, who must pass the application to the architect. The architect must ascertain (or cause the quantity surveyor to ascertain) the expenses and direct loss and/or damage caused to the employer by the determination and issue a certificate for the value of the work done and materials supplied, but not already certified. The employer is entitled to deduct his expenses and direct loss and/or damage when paying on the certificate provided he issues the requisite notices **[5.21]**.

The employer is also entitled to deduct the value of sub-contract work already paid for but found to be not in accordance with the sub-contract. The architect, of course, should never certify defective sub-contract work for payment. If he does, this provides a means of recovery. If the determination is because of the sub-contractor's failure to comply with the contractor's notice to remove defective work or correct defects in accordance with the sub-contract, the employer must take reasonable steps to recover the substituted sub-contractor's price from the original sub-contractor under the provisions of NSC/W. If he is not wholly successful, the contractor must pay the difference.

When making payments to the original sub-contractor after determination, the contractor may withhold 2.5% cash discount and the amount of his direct loss and damage.

4.12   Determination

Clause 35.25 makes clear that, notwithstanding the contractor's rights under the sub-contract, the main contract terms prevent him from determining the employment of a nominated sub-contract without an architect's instruction to that effect.

Clause 35.26 states that, if the sub-contractor's employment is determined due to the sub-contractor's default, the architect must give the contractor the information and direction in an interim certificate to enable him to specify the amount of loss and/or damage.

4.13   Nominated suppliers

Although clause 36.1.1 contains an elaborate and lengthy definition of a 'nominated supplier', it is first useful to look at what is *not* a nominated supplier. This information is contained in clause 36.1.2. If the source of materials has been specified in the bills of quantities and has been priced by the contractor, the supplier is not nominated even if there is only one source of the materials. In order for a supplier to be nominated, the materials must be the subject of a prime cost sum.

The detailed definition contains subtle traps for the unwary architect, particularly in regard to variation instructions and the expenditure of provisional sums. A nominated supplier can arise in one of the following ways:

- If there is a prime cost sum in the bills of quantities for materials and if the supplier is named in the bills or in the architect's instruction
- If there is a provisional sum in the bills of quantities for materials and if the supplier is named in an architect's instruction which makes the materials the subject of a prime cost sum or in a later instruction
- If there is a provisional sum in the bills of quantities and the architect specifies materials for which there is a sole source of supply, the architect must make them the subject of a prime cost sum and the sole supplier is deemed to be nominated
- If the architect instructs or subsequently sanctions a variation regarding materials for which there is a sole supplier, the architect must make them the subject of a prime cost sum in his instruction or written sanction and the sole supplier is deemed to be nominated.

It is easy to see that the creation of a nominated supplier may be inadvertent. However, it is possible to achieve much the same result in terms of naming a sole supplier if the provisions of clause 36.1.2, noted above, are followed.

There are two standard forms which may be used in connection with nominated suppliers: TNS/1 which is the form of tender, and TNS/2 which is a warranty in favour of the employer. Nothing in the contract places an obligation on either party or the architect to use these forms. However, it is probably true to say that such use is good practice.

Clause 36.3 sets out precisely how the amounts to be chargeable to the employer under clause 30.6.2.8 are to be calculated. The amount is to include any tax (other than VAT) or duty payable under Act of Parliament on the import, purchase, sale or treatment of the materials; net cost of packing, carriage and delivery; any price adjustment less discount other than cash discount for prompt payment; and any expense not reimbursed under clause 36.3.1.

The architect is only entitled to nominate a supplier who will enter into the terms of a contract of sale set out, at some length, in clause 36.4. There is a provision which allows the architect and contractor to agree to vary this provision. This provision is doubtless included to deal with the situation where a sole supplier of particular materials refuses to enter into a contract of sale on the terms specified. The terms are as follows:

4.13.1   Quality and standards: clause 36.4.1

The materials and goods must be as specified except so far as quality and standard are for the architect's approval when they must be to his reasonable satisfaction. This steps down to the supplier part of the contractor's responsibility under clause 2.1 with the same difficulties about the extent of the architect's opinion **[3.2]**.

4.13.2   Defects: clause 36.4.2

The nominated supplier is to make good defects in materials which appear up to the last day of the defects liability period and bear expenses incurred by the contractor provided that they are directly consequential. There are two provisos:

- The defects should not be capable of discovery by the contractor before fixing
- The defects must be solely due to defects in the materials and not to faulty storage or use by anyone for whom the sub-contractor is not responsible.

This clause deals with liability for defects becoming apparent up to practical completion, which would be implied in any event, and for defects after practical completion to the end of the defects liability period, which would not be implied. One of the provisos makes clear that the nominated supplier will not be liable if the contractor has failed to spot an obvious defect. Nothing in this term removes the supplier's liability for defects appearing and action taken against him up to the end of the limitation period.

4.13.3   Extension of the delivery period: clause 36.4.3

The supplier and the contractor are to agree a delivery programme which may be varied due to whichever of the following grounds they agree to include:

- *Force majeure*
- Civil commotion, strike or lock out
- Architect's instructions dealing with variations or provisional sums
- The architect's failure to supply information to, if required by, the supplier after specific application in writing at an appropriate time
- Exceptionally adverse weather conditions.

It is notable that the failure to provide information is stated to be to the supplier from the architect. Yet the contract makes no provision for the architect to provide the information directly in this way. Indeed, it would be remarkable if it did. It is all the more remarkable, therefore, that one of the grounds for varying the delivery programme refers to the architect's failure to do something which the contract does not require him to do in any event. The supplier is not a party to the main contract and the third party exclusion provision in clause 1.12 prevents any such obligation arising.

4.13.4   Cash discount: clause 36.4.4

The supplier must allow 5% cash discount for payments made within 30 days.

4.13.5   Determination: clause 36.4.5

The supplier need not deliver materials after determination of the contractor's employment unless payment has been made in full apart from cash discount.

4.13.6   Payment: clause 36.4.6

The contractor must pay in full, apart from cash discount, within 30 days after the end of the month in which the materials are delivered.

4.13.7   Vesting: clause 36.4.7

The contractor becomes owner of the materials when delivery is made to him or to someone else on the contractor's instructions. This is even if the supplier has not been paid.

4.13.8   Arbitration: clause 36.4.8

If a dispute between the supplier and the contractor is referred to arbitration, the provisions of the main contract arbitration clause 41B will apply.

4.13.9   Priority: clause 36.4.9

The provisions of clauses 36.4.1 to 36.4.9 inclusive take precedence over any other clause in the contract of sale.

Clauses 36.5.1 to 36.5.3 inclusive provide some protection to the contractor. They make clear that clause 36.5 is not to be interpreted as allowing the architect to nominate a supplier except in accordance with the terms set out in clause 36.4. Moreover, if there are any restrictions or exclusions of the supplier's liability in the contract of sale and the architect has specifically approved them in writing, the contractor's liability to the employer is similarly restricted.
   If the architect is dilatory in approving the restrictions, the contractor is entitled to wait until the architect has done so before entering into the contract of sale with the supplier.

4.14   Statutory requirements

The gas, electricity and water suppliers and other bodies regulated by statute including local authorities, may be involved either in performance of their statutory obligations or as contractors. The important thing to note is that while they are performing their statutory obligations, they do not enter into contracts. The earliest case in this connection appears to be *Milnes* v. *Huddersfield Corporation* (1886) where it was held that there was no contractual obligation by a water company created by Act of Parliament towards those it supplied. The same thing was said in relation to a gas company later in *Clegg Parkinson and Co* v. *Earby Gas Co* (1896): 'the obligation of the company if any, depends on statute and not upon contract'. The position is probably the same today although there are no longer water, gas or electricity boards.
   It is notable that in the case of *Read* v. *Croydon Corporation* (1938), Mr Justice Stable held that the relationship was not a contractual one, but 'a relationship between two persons under which one is

bound to supply water and the other, provided he has paid the equivalent rent, is entitled to receive it'.

This principle was apparently regarded as so axiomatic that none of these cases appears even to have been discussed in *Willmore* v. *South Eastern Electricity Board* (1957). Mr and Mrs Willmore started in business as poultry farmers rearing chicks by infra-red heat and the South Eastern Electricity Board promised them an adequate and constant supply of electricity to maintain lamps for that purpose. This the Board failed to do, with the result that the chicks died and Mr and Mrs Willmore were ruined financially. The judge held that the representations about the proper supply of electricity were not made 'with contractual intent' and there was no contract at all between the South Eastern Electricity Board and the unfortunate consumers: 'I have come to the conclusion that the plaintiffs, having failed to prove a contract, can have no cause of action of damages for breach of contract.'

Although they may not be liable for damages for breach of contract, that does not exonerate them in tort, e.g. for negligence.

JCT 98, therefore, draws a distinction between statutory undertakers performing their statutory duties and those situations where they act as contractors. Clause 6.3 states that clause 19 and 35 do not apply to statutory undertakers or local authorities carrying out part of the Works solely in accordance with their statutory obligations and they are not to be sub-contractors as referred to under the contract.

Every person or firm must comply with requirements laid down by statute. A contractor's duty to comply with statutory requirements will prevail over any express contractual obligation: *Street* v. *Sibbabridge Ltd* (1980).

Statutory obligations are dealt with in clause 6. The meaning of 'statutory requirements' is helpfully defined as being requirements of 'any Act of Parliament, any instrument, rule or order made under any Act of Parliament, or any regulation or by-law of any local authority or any statutory undertaker' relative to the works (JCT 98 clause 6.1.1). Statutory instruments are usually made by a secretary of state and the most important, from the contractor's point of view, will probably be the Building Regulations 1993 and the Construction (Design and Management) Regulations 1994, usually referred to as the CDM Regulations [4.15].

Under clause 6.1.1, the contractor must comply with all statutory requirements and give all notices which may be required by them. He is also responsible for paying all fees and charges legally demandable in respect of the Works. He is entitled to have such amounts added to the contract sum unless they are already provided for in the contract or unless they arise in respect of work or materials carried out or supplied by a local authority or statutory undertaker in a nominated capacity **[3.10]**.

Clause 6.2 contains an indemnity provision which deserves attention. Not only must the contractor pay charges as above, but he agrees to indemnify the employer against liability in respect of such charges. Therefore, if the contractor fails to pay as legally required, he assumes liability on behalf of the employer. Such liability might well extend to undoing work already done, delays or fines. This is an onerous provision which might easily be overlooked. Its purpose is to keep the employer safe from damage or loss. Although indemnity clauses tend to be interpreted by the courts against the person relying on them, it is not thought that this will give the contractor much comfort here. In addition, the time during which he remains liable under an indemnity clause does not begin to run until the liability of the employer has been established, usually by the court.

Clause 6.1.2 provides that the contractor has a duty to notify the architect immediately if he finds a divergence between statutory requirements and the contract documents or architect's instructions. It seems that the contractor has no obligation to search for such divergencies: *London Borough of Merton* v. *Stanley Hugh Leach Ltd* (1985). When the architect receives the notice, or if the architect finds a divergence himself, clause 6.1.3 gives him just seven days to issue an instruction about the divergence. If the architect fails to meet this deadline, the contractor would appear to have clear grounds for an extension of time under clause 25.4.6.2 and reimbursement of loss and/or expense under clause 26.2.1.2. Although the clause does not expressly say so, it must be implied that the architect's instructions are to remove the divergence. If the work is varied as a result, the architect's instruction is to be treated as though it was issued under clause 13.2.

Provided that the contractor notifies the architect if he finds a divergence and otherwise carries out the work in accordance with the contract documents and any other drawings and instructions issued by the architect, the contractor is given a valuable safeguard. He is not to be held liable to the employer if the Works do not comply with statutory requirements. The contractor is still liable for compliance as far as the local authority is concerned and he is still obliged to comply with statute, but if obliged to rectify non-complying work, he should be able to recover his costs from the employer. Thus, if the contractor fails to find a divergence, he is able to escape liability to the employer by virtue of this clause. An interesting situation would arise if a contractor finds a divergence, notifies the architect and the architect refuses to issue an instruction to deal with the problem. Although such an occurrence must be very rare, if it happened the contractor would probably be obliged to refuse to knowingly build in contravention of statutory requirements **[3.4]**.

The position is significantly amended if performance specified work is involved **[3.13]**. Clauses 6.1.6 and 6.1.7 deal with it. If a divergence is found between statutory requirements and the contractor's statement, the finder must immediately notify the other in writing, but it is the contractor who must propose an amendment to deal with the divergence. The architect is still required to issue an instruction to put the amendment into effect and to make it part of the Works. Presumably, the architect's instruction will generally echo the contractor's proposal. However, if the architect's instruction does not faithfully represent the contractor's proposal and the contractor believes that it will 'injuriously affect' the Works (to use the contract's own quaint expression) the contractor is able to put the clause 42.15 procedures into operation **[3.13]**. Otherwise, the contractor is not entitled to any addition to the contract sum unless the root of the problem is a change in statutory requirements which took place after the base date; in which case the amendment is treated as though it was an architect's instruction under clause 13.2 requiring a variation and the result is valued accordingly.

Contractors are not entitled to carry out instructions from anyone other than the architect. It is not unknown for a building control officer to visit site and to direct the contractor that work does not comply with the Building Regulations. However tempting it may be for the contractor simply to comply with the direction, particularly if it seems that there is no possible alternative, the contractor must do nothing until he has referred the matter immediately to the architect and received an instruction. The only exception to this is in the case of an emergency. Clause 6.1.4 provides that if the matter really is urgent, the contractor may carry out the necessary work immediately provided that he notifies the architect forthwith of the steps he is taking, and supplies only sufficient materials and carries out just enough work to ensure compliance. The contractor is then entitled to payment for what he has done just as though the architect had issued an instruction.

4.15   *The CDM Regulations 1994: clause 6A*

Breach of the Construction (Design and Management) Regulations 1994 (1995 in Northern Ireland) will be a criminal offence, but except for two instances, a breach will not give rise to civil liability. Thus one person cannot normally sue another for breach of the Regulations. Compliance with the Regulations is made a contractual duty so that breach of the Regulations is also a breach of contract. This is likely to cause problems for the employer.

Article 6.1 assumes that the architect will be the planning supervisor. The word 'or', enables the user to insert an alternative name. Article 6.2 defines the principal contractor as the contractor. Article 6.1 is not sufficient to bind the architect to the employer for the purpose of carrying out the function of planning supervisor and a separate contract for these services should be executed.

There are sundry definitions and words which make clear that the Works must be carried out in accordance with the health and safety plan. Grounds for determination (failure to comply with the Regulations) are included in the list in both employer and contractor determination clauses (clauses 27 and 28).

Clause 6A.1 provides that the employer 'shall ensure' that the planning supervisor will carry out all his duties under the contract and, where the principal contractor is not the contractor, he will carry out his duties in accordance with the Regulations. This is little short of a guarantee whether the employer realises it or not. There are also provisions that if the contractor is the principal contractor, he will comply with the Regulations (clause 6A.2). The contractor must also ensure that any sub-contractor provides necessary information. Compliance or non-compliance by the employer with clause 6A.1 is a 'relevant event' and a 'matter' under clauses 25 and 26 respectively. Lest the significance is missed, what this means is that the employer must ensure that the planning supervisor and the principal contractor, where he is not the contractor, perform correctly and if they do not, or even if they do, any resultant delay or disruption will give entitlement to extension of time and loss and/ or expense. This may well be a most fruitful source of claims for contractors. Every instruction potentially carries a health and safety implication which should be addressed under the Regulations. The regulations impose substantial duties on the planning supervisor. Most of them are to be found in Regulations 14 and 15. Some of these duties must be carried out before work is commenced on site, which may present difficulties where the contractor is not appointed until comparatively late in the process. If necessary actions delay the issue of an instruction or, once issued, they delay its execution, the contractor may be able to claim.

There may be rare occasions when the Regulations do not fully apply to the Works as described in the contract. If the situation changes due to the issue of an instruction or some other cause, the employer may be faced with substantial delay as appointments of planning supervisor and principal contractor are made and appropriate duties are carried out under the full Regulations.

It is certain that the key factor will be for employers, planning supervisors and principal contractors to structure their administrative procedures very carefully if they are to avoid becoming in breach of their contractual obligations in regard to the CDM Regulations.

4.16   *Work not forming part of the contract*

The employer has the right to make contracts with persons other than the contractor to carry out work on the site. The employer is not entitled to deduct work from the contractor so as to give it to another contractor. That would be a breach of contract which is certainly not contemplated by either clause 13.1.1.1 or clause 29: *Vonlynn Holdings Ltd* v. *Patrick Flaherty Contracts Ltd* (1988); *AMEC Building Contracts Ltd* v. *Cadmus Investment Co Ltd* (1997). Clause 29 provides for two situations:

- The bills of quantities may provide the contractor with very full information so that he can properly carry out the work required of him under the contract. They may also note that specific work is not to form part of the contract and will be carried out by others. In such an instance, the contractor must permit the specific work.
- The bills of quantities may not provide the full information noted above. In that case, the employer may still employ other persons to do work not included in the contract provided that the contractor gives his consent. The contractor must not unreasonably withhold his consent.

Delays caused by persons engaged by the employer under clause 29 may give rise to an extension of time for the contractor under clause 25.4.8 or the payment of direct loss and/or expense under clause 26.2.4. The delays may be caused by persons failing to carry out the work, carrying it out slowly or simply carrying it out properly. For the employer to engage other contractors is like signing a blank cheque.

Persons engaged by the employer under this clause are deemed to be persons for which the employer is responsible for the purposes of the indemnity clause (clause 20) and not a sub-contractor **[8.2]**.

CHAPTER FIVE
CERTIFICATES, PAYMENTS AND RETENTION

5.1 The architect as certifier

In certifying, the architect must act fairly and impartially between the parties: *Hickman & Co* v. *Roberts* (1913) and *Panamena Europa Navigacion* v. *Frederick Leyland & Co Ltd* (1947). It was once said to be 'established law' (by Lord Radcliffe in the House of Lords in *R.B. Burden Ltd* v. *Swansea Corporation* (1957)) that the architect acted in an 'arbitral capacity'. As a result, the Court of Appeal held that a certifying architect was immune from actions for negligence by either the employer or the contractor.

That 'established law' was swept aside by the decision of the House of Lords in *Sutcliffe* v. *Thackrah* (1974). An architect had given an interim certificate of £3090 on which the builder had been paid. Part of the work included in the certificate was defective. Normally, such an over-certification would have been adjusted by the next certificate since each certificate certifies a cumulative value, but in this case the contractor became insolvent and the money overpaid was irrecoverable. On a preliminary point of law it was held that an architect who was called on to certify the value of work done for an interim certificate under any building contract was liable in both contract and tort to his employer for negligent over-certification. In passing, it would, therefore, appear that he may also be liable in tort to a contractor for negligent under-certification: *Stevenson* v. *Watson* (1879) **[2.7]**.

Sutcliffe v. *Thackrah* began life as *Sutcliffe* v. *Chippendale and Edmondson* (1971) in the Official Referee's court. It is worth looking at what Judge Stabb had to say. The House of Lords did not subsequently disturb this opinion. The judge was concerned with whether the quasi-arbitrator privilege, which was then considered to protect the architect in giving the final certificate, extended to protect him when issuing interim certificates. He quoted Mr Justice Sankey in *Wisbech Rural District Council* v. *Ward* (1927):

'Although it is probably right to say that in giving a final certificate the architect acts in a quasi-judicial character unless there is
some express clause in the contract to contradict it, it cannot, I
think, be asserted that in giving an interim certificate he is so
acting. Personally, I should have thought that the inference was
just the other way – namely, that in giving an interim certificate
he is merely acting as an agent for the building owner unless
there is something in the contract to contradict that.'

Judge Stabb said:

'Plainly, it is part of an architect's duty of supervision, as agent for
the employer, to see that the work is properly executed, and
therefore to my mind supervision and the issuing of interim
certificates cannot be regarded as wholly separate and distinct
functions.

I think it was rightly contended on behalf of the plaintiff that in
a well supervised contract an architect would not certify for work
not properly executed.

I have come to the conclusion that the architect, in discharging
his function of issuing interim certificates, is primarily acting in
the protection of his employer's interests, by determining what
payment can properly be made on account, such determination
being based upon his assessment of the value of the work
properly carried out, that assessment being performed by virtue
of his professional skill, for which the building owner has
engaged him.

It is in my view part of his supervisory function to see that the
value of work properly executed, and only work properly executed, is included in the valuation for the purpose of an interim
certificate, and that therefore he is under a duty of care to his
employer in the performance of that function.

Accordingly, if by his failure to exercise due skill and care, he
should fail to exclude from the certificate the value of work not
properly executed, he would in my view be liable to his employer
for any damage attributable to such default.'

He also took the view that the architects in this particular case:

'must have known at least of the possibility that these contractors
would not be recalled to remedy those defects. They knew that
the contract had been nearing completion.

In these circumstances, I consider that it was their plain duty to be particularly accurate in their valuation of the work properly executed to that date.'

The standard of care required from professional persons is to use the reasonable skill and care of such persons of ordinary competence measured by the professional standard of the time. This is commonly referred to as the *Bolam* standard from *Bolam* v. *Friern Hospital* (1957) in which it was formulated for the guidance of a jury. Dealing with the standard applicable to an independent valuer in *Belvedere Motors Ltd* v. *King* (1981) Mr Justice Kenneth Jones said:

> 'The defendant can only be found guilty of negligence if it can be shown that he has omitted to consider some matter which he ought to have considered, or that he has taken into account some matter which he ought not have taken into account or in some way has failed to adopt the procedure and practices accepted as standard in his profession, and has failed to exercise the care and skill which he, on accepting the appointment, held himself out as possessing.'

It is evident, therefore, that the architect must be liable to his employer if he issues a certificate to the contractor for payment of on-site goods without satisfying himself that the contractor can pass such a title to the employer as will make the employer the owner of the goods for which he has to pay.

It seems no longer sufficient for the architect to say that a substantial body of architects would have acted in the same way. That argument will collapse if the action can be shown to be unreasonable, illogical or irresponsible: *J.D. Williams & Co Ltd* v. *Michael Hyde & Associates* (2001).

In *Ashwell Scott Ltd* v. *Scientific Control Systems Ltd, Milton Keynes Development Corporation (third parties)* (1979), it emerged in the course of the hearing that there was an ingenious arrangement between all the parties whereby the air conditioning to be installed was leased from Eastlease Ltd, the leasing subsidiary of the Norwich Union, with the benefit of 100% write-off against corporation taxation as 'plant', while at the same time constituting 'new build' and, therefore, zero rated for the purposes of VAT. At the same time, the engineer was to include it in certificates he issued to the defendants (Scicon) for payment to the contractors. It was in connection with this that the judge made the remark: 'An engineer who includes in a certificate goods which are not the property of the contractor does so at his own peril'.

In the particular circumstances of this case, no doubt the employer would be estopped from complaining that the engineer, in issuing certificates for goods the title of which he well knew was not intended to pass to his employer, was negligent, and the judge held that title *had* in fact passed when the air conditioning became incorporated in the property.

The architect who included the slates in his interim certificate in *Dawber Williamson* v. *Humberside County Council* (1979) could possibly have a similar defence against his employer – namely that the employer, by entering into JCT 63, was estopped from claiming that his architect was negligent in certifying sums for goods of which the employer could not become the owner.

It is clear, however, that an architect may be liable if he includes in interim certificates the value of defective work or unfixed materials for which the contractor has no title.

5.2   Interim certificates: valuation

JCT 98 clause 30.1.1.1 requires the architect to issue interim certificates to the contractor at the intervals stated in the appendix. If none is stated, the interim certificates are to be issued at intervals of one month. The amounts to be included in those certificates are set out in Table 5.1.

The provision in clause 30.2.1.2 for the inclusion of on-site materials and goods raises difficulties. The goods may have been delivered to the contractor subject to a retention of title clause by the supplier. In that event, ownership of the goods will only pass to the contractor if he has paid for them, or to the employer when they become permanently affixed to the property. Or the unfixed materials may belong to a work and materials sub-contractor, in which case it will never be the intention of the parties that the contractor should have a title to pass to the employer under clause 16.1, so that the situation which arose in *Dawber Williamson* can easily recur [5.3].

5.3   Retention of title on the sale of goods

Retention of title clauses, which are expressly recognised by section 19(1) of the Sale of Goods Act 1979, have been recognised as valid ever since the House of Lords case of *McEntire and Maconchy* v. *Crossley Brothers Ltd* in 1895. The full value of such a clause,

TABLE 5.1
Amounts to be included in interim certificates:
Clause 30.2

Add:
Subject to retention:

- Value of work properly executed
- Value of materials and goods if delivered to site at the appropriate time
- Value of listed off-site materials
- Amounts for each nominated sub-contractor in accordance with NSC/C clause 4.17.1
- Contractor's profit on NSC/C clauses 4.17.1 and 4.17.2.

Not subject to retention:

- Payments in connection with statutory fees and charges, opened up work in accordance with the contract, royalties, employer's non negligence insurance, cost of premiums as a result of the employer's failure to insure under clauses 22B or 22C
- Loss and/or expense, or restoration costs in connection with clauses 22B or 22C
- Early final payment of nominated sub-contractors
- Fluctuation payments under clauses 38 or 39
- Amounts for each nominated sub-contractor in accordance with NSC/C clause 4.17.2.

Deduct:
Not subject to retention:

- Amounts to be deducted following incorrect setting out, defective work, defects during or at the end of the defects liability period
- Amounts for each nominated sub-contractor in accordance with NSC/C clause 4.17.3.

From this total is to be deducted:

- The appropriate retention
- The amount of any due reimbursement of an advance payment
- The total amount previously certified as due.

however, did not occur to manufacturers and suppliers until the case of *Aluminium Industrie Vaassen* v. *Romalpa* in 1976 . Since then, they have been widely used, not least by manufacturers of building products, who have learnt by bitter experience that if there is any insolvency in connection with a building contract, it is they who suffer most of all as unsecured creditors, rarely receiving any dividend in the liquidation even though their materials may have been incorporated into the fabric of the building and are of value to the owner.

The form of retention clause varies. A simple retention of title clause may read: 'The property in all goods is not to pass to the customer until they are paid for in full'. A 'current account' clause may retain property until all indebtedness by the customer in respect of all transactions has been discharged. A 'proceeds of sale' clause may seek to make monies received for the goods the property of the vendor, and an 'aggregation' clause may purport to give ownership of any goods admixed with or joined to the vendor's goods to the vendor.

Such clauses will, in most cases, be defeated when the goods are affixed to the realty and cease to exist as chattels. However, it is possible by contract to provide a right for an unpaid vendor to sever such chattels from the realty.

In *Hobson* v. *Gorringe* (1897) it was held that the owner of goods supplied on hire purchase was entitled to enter on the land and remove his goods which had been affixed to the property. The terms of the hire purchase agreement constituted a licence, coupled with an interest in the land. Such a clause may be effective against the other contracting party, i.e. the contractor in possession of the site and his liquidator or receiver, but will rarely defeat debenture holders with floating charges or mortgagees of specific property with charges created before the addition of the chattel to the property: *Longbottom* v. *Berry* (1869); *Meux* v. *Jacobs* (1875). If the charges are created after affixation, it would appear that the right to enter and recover the goods, even affixed goods, takes priority over subsequently created equitable interests, whether the debenture holders are aware of the hire purchase agreement or not: *Re Morrison Jones & Taylor Ltd* (1914).

Therefore, it may be open to the supplier to write a clause such that, even where goods such as doors have been affixed to the property, an unpaid vendor would have the right to remove them in the event of the contractor's insolvency.

5.4   *Interim certificates: off-site materials and goods*

Amounts included in interim certificates which are subject to deduction of retention monies are set out in Table 5.1. There has always been a problem with certification for payment of off-site materials and goods. The problem is generally that it is difficult to be sure that the materials and goods, once paid for, will eventually be delivered to site. The main danger is insolvency of the contractor. The architect has virtually no discretion as regards the certification under clause 30.3. The scheme of the clause is simple in essence, but complex enough in expression to deter its operation in practice. Anecdotal evidence suggests that payment for off-site materials is very much the exception in recent contracts.

If the employer wishes to pay for off-site materials, he must make a list of them (referred to as 'the listed items') and supply the list to the contractor and attach the list to the contract bills. Clearly the contractor will have to be supplied with the list at time of tender, because whether or not he will be entitled to payment for off-site materials will be an important factor in his pricing strategy.

Clauses 30.3.1 and 30.3.2 refer to two types of listed items: 'uniquely identified items' and 'not uniquely identified items' respectively. The terms are not defined anywhere in the contract although the JCT guidance notes give its version. The guidance notes are not part of the contract, however, and it does not matter what JCT thinks it means or indeed intended it to mean. Being undefined, the terms must be given their ordinary meaning and that immediately causes a problem. 'Unique' means the only one of its kind. So does 'uniquely identified' mean an item which is the only one of its kind or an item which has been identified in a way which cannot be repeated? Although the JCT guidance notes state that 'uniquely identified items' refers to such things as a boiler and that 'not uniquely identified items' are things such as a stack of bricks, it is by no means certain that this is the proper meaning. No doubt in due course a court may be invited to arrive at a definition and it may or may not adopt the JCT view.

Fortunately, very little seems to hang on the point. In the case of uniquely identified items the contractor may, and in the case of items not uniquely identified items the contractor must, provide a bond in the form agreed between the JCT and the British Bankers' Association and annexed to the appendix. In each case, the surety must be approved by the employer. It should be noted that the seventh recital refers to the situation where the employer desires a bond in different terms. In that case, the contractor must have been given a copy of the terms (and of course agreed them) before executing the contract. Indeed, it is advisable to let the contractor have a copy at tender stage along with anything else which the employer proposes to change from the standard. Details of the amount of the bond are to be inserted in the appendix. Although 'unique' and 'not unique' are differentiated, it is perfectly possible to delete them and insert reference to one bond for all the items.

The listed items, whether or not they are uniquely identified, are to be included in interim certificates provided that certain criteria are satisfied:

(1)   The need for a bond – as discussed above.

(2)   The contractor must have provided the architect with reasonable proof that the contractor owns each of the items. Proof of ownership, reasonable or otherwise, is not easy to achieve. The JCT guidance notes suggest some things which the contractor may do and which may satisfy the architect: copies of contracts of sale, statements from the suppliers that the contractor has satisfied pre-conditions and copies of sub-contract terms with appropriate clauses. The problem is that the architect requires some legal expertise to be able to interpret these documents and the sensible architect will insist that the employer takes proper independent legal advice about such matters. Clause 16.2 then provides that after the value has been included in an interim certificate and paid by the employer, the items become the property of the employer. The contractor is not to remove the items from their storage except to take them to site. The contractor remains responsible for any loss or damage until the items are 'on or adjacent to the Works'.

(3)   The items must be in accordance with the contract. This is somewhat otiose, because the architect, of course, has no power to certify anything which does not comply with the contract.

(4)   The items must be set apart from other items or they must be clearly marked in some way so as to identify the employer and the Works as their destination. No doubt this is a good idea, but it would have little force against a devious contractor. Setting aside and marking cannot be a permanent state, because the items are destined for the Works where any such marks must be capable of removal.

(5)   Reasonable proof must be given to the employer (not the architect) that the items are insured for full value against loss or damage against specified perils. The policy must protect the

interests of both employer and contractor and therefore, although it is not expressly stated, the policy should be in joint names.

5.5   *Interim certificates: work 'properly executed'*

The major part of the content of an interim certificate is likely to be 'work properly executed' (clause 30.2.1.1). It is important to understand exactly what this expression means. Fortunately, it has been the subject of careful judicial consideration which is still relevant today. In *Sutcliffe* v. *Chippendale and Edmonson* (1971) the judge had the benefit of expert witnesses:

'I have had the advantage of ... listening to the views of no less than five experienced and independent architects and quantity surveyors ... in addition to the two architects and two quantity surveyors involved in the case.

It is clear that there is a variation within the profession as to the practice in the preparation of the certificate, which is only to be expected, and also some divergence of principle as to what the amount certified is designed to represent, which is perhaps more surprising.'

One architect said that it was his practice, as a matter of routine, when he paid a site visit if he observed defective work to issue a written order to the contractor, with a copy to the quantity surveyor so that he would know not to include it in his next certificate. Without such a system, he said, the architect would have to get in touch with the quantity surveyor in some other way to pass on the information. The judge said of this witness:

'He took the view that the certificate should conform as strictly as possible to the terms of the contract and that, if he was not satisfied with any particular item of work, it was implicit that the value of the work was not to be included in a certificate covering work properly executed.'

Other witnesses, the judge said, expressed the view:

'that such a certificate was more of an approximation of the value of the work as it progressed, assessed by the quantity surveyor without any detailed inspection of the work, the object being

simply to provide a reasonable progress payment for the contractor based upon a comparatively cursory examination of the
site.

All were agreed that responsibility for the detection and, if
necessary, the exclusion from the certificate of defective work was
that of the architect as opposed to the quantity surveyor, whose
concern was as to quantity and not quality.'

The judge concluded:

'Faced with these two opposing views held by men of experience
in the profession, my task of deciding what is an architect's
proper professional function is plainly a difficult one.

Since everyone agreed that the quality of the work was always
the responsibility of the architect and never that of the quantity
surveyor and since work properly executed is work for which a
progress payment is being recommended, I think that the architect is in duty bound to notify the quantity surveyor in advance of
any work which he, the architect, classifies as not properly executed, so as to give the quantity surveyor the opportunity of
excluding it.

As to the system or method of communication between architect and quantity surveyor to be adopted, I make no comment,
save that for a busy architect merely to rely upon his memory for
this purpose seems, in my view, to be unsatisfactory.

Furthermore, I can well understand an architect's reluctance to
devote too much time to matters concerned with certificates at
interim stage, since, in the normal course of events, work that
may be defective and unacceptable at that stage, can and will be
remedied at a later stage, and therefore overcertification and
consequent overpayment would only be temporary and must
automatically adjust itself as the contract reaches completion, and
defects are remedied by the contractor at his own cost before any
final certificate is issued.

But should circumstances unexpectedly bring the contract to a
premature end whilst such overpayment remains uncorrected, it
is difficult to see how the architect can avoid responsibility, if the
payment proves to be irrecoverable from the contractor as has
happened in this case.

In my considered opinion the strict approach is the right one.

If meaning is to be given to the words "work properly executed" in clause 30 sub-clause 2 [of JCT 63], I cannot see how the
architect can avoid the requirement to exclude work, which is not

properly executed, from the value of the work for which he recommends his employer to make payment on account, and if the work is defective and unacceptable as it stands, it must, in my view, be classified as work not properly executed until the defect has been remedied.

I do not accept that the words "work properly executed" can include work not then properly executed but which it is expected, however confidently, that the contractor will remedy in due course.

So long as the contractual basis of the certificate is the valuation of the work *properly* executed, the architect should first satisfy himself as to the acceptable quality of the work before requiring his employer by way of certificate to make payment for it, and in particular should keep the quantity surveyor continually informed of any defective or improperly executed work which he observed.'

With respect, that is a superb exposition of the architect's position. JCT 98 clause 30.10 specifically provides that the contractor cannot rely on any certificate, other than the final certificate in certain circumstances **[5.15]**, as conclusive evidence that any work, materials, goods or performance specified work to which it relates are in accordance with the contract.

If a certificate is wrongfully withheld or the employer's breach of contract prevents its issue, the contractor can take action to recover the money without the certificate: *Panamena Europa Navigacion* v. *Frederick Leyland & Co* (1947) and followed in *Croudace Construction Ltd* v. *London Borough of Lambeth* (1986).

5.6   Retention monies: the amount

Before considering the clauses which allow the retention of monies, it should be noted that clause 30.4A has been introduced into the contract to give the parties the option of a bond from the contractor instead of retention. If the appendix states that a bond is to apply, the contract provisions allowing deduction of retention and its subsequent release do not apply, but the architect, or the quantity surveyor if so instructed, must prepare a statement showing the amount which would otherwise have been deducted from the contractor and each nominated sub-contractor (clause 30.4A.1). If the contractor defaults in the provision of a bond, the retention provisions take effect from the next interim certificate and any such retention deducted is not released until the breach is rectified and the bond is properly in place.

Clause 30.4A.2 requires the bond (on terms set out in the contract) to be provided to the employer no later than the date of possession.

In a somewhat complex provision (clause 30.4A.4), if the bonded sum falls below what a retention fund would have held, the contractor must either take steps to increase the bond or appropriate amounts of retention may be deducted to make up the balance.

Clause 30.4A.5 makes clear that if there is also a separate performance bond and a claim could be made by the employer on either bond, the claim must first be made on the retention bond.

If a bond is not used, certain of the amounts to be included in the certificate are subject to deduction of retention (see Table 5.1). The purpose of this is to provide a fund from which the employer is entitled to draw money in certain specified instances. The right to take money from the retention fund is enshrined in clause 30.1.1.2. This is a clause referring to rights contained in the contract of withholding or deduction against monies due or to become due to the contractor. It makes clear that the right of deduction exists despite the retention being held in a trust fund **[5.7]**. It also appears that deduction from retention can take place even if there is no retention amount due to the contractor under the interim certificate from which it is proposed to make such deduction. It is clear that the employer has no general power to deduct from the retention fund. The right is subject to the operation of clause 35.13.5.3.2 where nominated sub-contractors are concerned **[4.9]**. If the employer does properly deduct money from the retention fund, clause 30.5.4 states that he must notify the contractor of the amount and whether it is from the contractor's or the nominated sub-contractor's retention as set out in the interim statements referred to in clause 30.5.2.1.

The rules for ascertaining the amount of retention are contained in clause 30.4. Clause 30.4.1.1 provides that the percentage is to be 5% unless the parties agree a lower percentage and it is inserted in the appendix. A footnote suggests that if at tender stage, the employer estimates the contract sum to be over £500,000, the retention should be no more than 3%. The footnote does not create any contractual obligation in the absence of a specific agreement on this point and the insertion of a lower rate in the appendix. A percentage higher than 5% can be inserted in the appendix for the retention if the contractor agrees. It would be unusual except on very small projects and the tendering contractors would no doubt increase their tender figures accordingly. Sometimes, an employer wishes to retain money at the rate of 5% until the fund reaches 3% of the contract sum and then continues at 3% thereafter. That is a way of reaching the total retention earlier than would otherwise be the case – a kind of front loading on the employer's part. If such an arrangement is desired, it is necessary to draft an appropriate additional clause to cover the point.

The scheme of the retention clause is that the full percentage is to be deducted until practical completion is certified **[12.1]**, after which half the retention is released. The second half of the retention is not released until the certificate of completion of making good defects has been issued **[12.4]**.

5.7   Retention monies: a trust fund

Clause 30.5 deals with the treatment of the retention fund. The employer's interest is described as 'fiduciary as trustee'. That applies to the contractor and to any nominated sub-contractor. The essence of clause 30.5.1 is that the retention money is the contractor's or the nominated sub-contractors' money. The employer does not own it. He has no legal or equitable interest in it. He merely holds it as trustee. The contractor and nominated sub-contractors are beneficiaries. Important implications of law follow from this.

The intention, of course, is to create a situation such that if the employer becomes insolvent, retention monies in his hands are not his property: they are branded with the proprietary interests of the contractor or sub-contractor beneficiary and a receiver or liquidator or trustee in bankruptcy cannot lay his hands on them.

But once a trust is created, certain effects inevitably follow.

Perhaps the most important is that the trustee is obliged to invest trust monies in such investments as are prescribed by the Trustee Investment Act 1961. The words of JCT 98 clause 30.5.1 'but without obligation to invest' purport to exonerate the employer from the obligations of that Act and of the Trustee Act 1925.

5.8   Separate fund necessary for retention monies

A trustee is under an obligation not to mingle trust monies with his own. This means that he must establish a separate trustee account. In *Rayack Construction Ltd* v. *Lampeter Meat Co Ltd* (1979) Rayack entered into a contract in 1978 in the JCT 63 form which contained clause 30(4) in the same terms as JCT 98 clause 30.5.1. They sought a declaration that the employer (Lampeter) was obliged to pay the retained monies into a separate bank account to be applied only in accordance with the trust specified in clause 30(4) (now 30.5.1).

The employer claimed that the clause contained no express provision requiring them to establish a separate fund. The court accepted that the clause did require the employer to establish a separate fund. Mr Justice Vinelott said:

'In my judgment, clause 30(4) construed in the context of the agreement as a whole, does impose an obligation on an employer to appropriate and set aside as a separate trust fund retention monies.

Unless clause 30(4) is construed as imposing such an obligation it cannot have any practical operation.

Further, clause 30(4) refers to the 'contractor's beneficial interest therein' and the predicated beneficial interest could only subsist in a fund so appropriated and set aside.

It is in my judgment clear that the purpose of retention under clause 30(4) is to protect both employer and contractor against the insolvency of either.

[Counsel for the defendants] says that they might be faced with a cash flow problem if they were now compelled to appropriate and set aside these monies. That argument is double-edged.

It would be wrong to expose the plaintiffs to any degree of jeopardy in order that the defendants might continue to use in their business monies which ought to have been appropriated and set aside for the security of the plaintiffs.'

In other words, in all circumstances, contractors are entitled to have separate accounts set up in respect of retention monies as trust funds. If this is not done, it is possible that equity will treat as done that which ought to be done. In *Re Arthur Sanders Ltd* (1981), Mr Justice Nourse in a case on JCT 63 and the NFBTE/FASS Green Form held that:

'Once the sums notionally set aside have been impressed with the relevant trusts, they remain subject to those trusts, whatever the fate of the contractor's employment or of the contract itself.'

Similarly it followed that where JCT 63 is entered into between the contractor and the employer and the NFBTE/FASS form between the contractor and the sub-contractor, 'it creates a trust in favour of the sub-contractor'.

The editors of *Building Law Reports* suggest in a note to the *Rayack*

case that since all injunctions, mandatory or prohibitive, are dis-
cretionary, another court might in different circumstances not order
separate funds for retention monies. But no court of equity could by
exercise of discretion exonerate any trustee from the fundamental
obligations of being a trustee, even with the consent of the beneficiaries.

The definitive word on the matter was given by the Court of
Appeal in *Wates Construction (London) Ltd* v. *Franthom Property*
(1991). The case concerned a contract under the former JCT design
and build contract, CD 81. The retention clause, 30.5.1, is virtually
identical to that in JCT 98. Clause 30.5.3 in both contracts requires
the employer to place the retention monies in a separate bank
account at the date of payment of each certificate if the contractor or
any nominated sub-contractor so requests. The employer is
required to certify the deposit to the architect with a copy to the
contractor. In the *Wates* case clause 30.5.3 had been deleted, but the
contractor applied to have the retention monies deposited in a
separate bank account. Lord Justice Beldam said:

> '... clause 30.5.1 creates a clear trust in favour of the contractors
> and sub-contractors of the retention fund of which the employer
> is a trustee. The employer would be in breach of trust if he
> hazarded the fund by using it in his business and it is his first
> duty to safeguard the fund in the interests of the beneficiaries ...'

The deletion of clause 30.5.3 caused him no difficulty:

> 'Firstly, it seems to me that there is no ambiguity about the part of
> the agreement which remains. The words of clause 30.5.1 under
> which the trust is created are quite clear. Secondly, the fact of
> deletion in the present case is of no assistance because the parties,
> in agreeing to the deletion of clause 30.5.3, may well have had
> different reasons for doing so and it is not possible to draw from
> the deletion of that clause a settled intention of the parties com-
> mon to each of them that the ordinary incidence of the duties of
> trustees clearly indicated by clause 30.5.1 were to be modified or
> indeed removed. It may have been thought by one of the parties
> to have been unnecessary to have included clause 30.5.3. It may
> have been that one of them thought that the employer should
> have been liable to account for any interest to the contractor if the
> retention fund was placed in a separate account. But there may be
> various reasons, which it is not possible to set out in full, why the
> clause was deleted and it is quite impossible to draw any clear

inference from the fact of the deletion. I therefore would reject an argument based upon the fact of deletion and can see no ambiguity upon which reference to that deleted clause could assist.'

It seems, therefore, that the employer has an obligation to put retention monies in a separate bank account whether or not the contractor so requests and whether or not there is a clause to that effect. In any event, the contractor certainly does not have to make more than one request: *Concorde Construction Co Ltd* v. *Colgan Co Ltd* (1984). It should be noted that the local authorities edition of JCT 98 does not include clause 30.5.3. The words 'Number not used' are inserted. Following the *Rayack* and *Wates* cases, it appears that a local authority acting as employer would still be obliged to set up a separate bank account for the retention trust fund.

The setting aside of retention monies is, therefore, extremely important to the contractor and determination of his employment has no effect on his right to request the fund to be set aside. It seems, however, that the trust is not established until the fund is set aside and if the employer becomes insolvent before the setting aside has taken place, the contractor will be unable to exert a better claim over the retention than any other unsecured creditor: *MacJordan Construction Ltd* v. *Brookmount Erostin Ltd (in Administrative Receivership)* (1991).

Although all beneficiaries can in some circumstances alter the terms of a trust by deed, they can never alter the fundamental obligations of trustees established over the years by the courts of equity.

5.9   Is the contractor entitled to interest on retention?

Trustees must never benefit from trust monies. A trustee is not entitled to remuneration from a trust. Only a solicitor trustee with a charging clause in the trust may under the rule in *Cradock* v. *Piper* (1850), if he is acting for himself and other trustees, charge the costs of so acting; and then only if his costs would have been no greater than if he was acting for the lay trustees alone.

It follows from this that no trustee is ever entitled to profit from his position of trust. In one case, where the trustees of an estate went shooting grouse on it, they were held liable to account to the beneficiaries for the value of the shooting. A stockbroker's clerk who was trustee of an estate was not entitled to commission from his employers when they were asked to value it. The test is not whether the trust has suffered loss by the trustee's action. It is whether an individual has benefited by his position of trustee.

In *Swain* v. *Law Society* (1981) the Law Society, being in a fiduciary position, was held liable to repay to individual members commission it had received from the insurance brokers on its members' compulsory insurance. There is no escape from this rule that trustees or those in fiduciary positions analogous to trustees must not benefit from the trust.

As Lord Justice Stephenson said in the Court of Appeal in the *Swain* case:

> 'The Law Society obtained the insurance contract by reason of its fiduciary position ... The commission was acquired by the Society using its fiduciary position. It is therefore accountable for it to the solicitor beneficiaries.
>
> There is *an inflexible rule of equity*, exemplified in *Phipps* v. *Boardman* (1966) ... that a person who owes a fiduciary duty to others is accountable to them for any profit or other advantage he obtains from his fiduciary position.' (author's emphasis)

Although the House of Lords subsequently set aside this judgment, it did not dispute the description of the obligations of trustees but used the ground that the Law Society, in setting up the insurance scheme, was exercising statutory powers and was not in the position of ordinary trustees.

There is an exact analogy between that and employers in the position of contractor's retention monies. 'An indelible incident of trust property is that a trustee can never make use of it for his own benefit,' said Lord Cottenham in *Foley* v. *Hill* as long ago as 1848. He cannot do so even if the beneficiary consents. There can be no doubt that under JCT 63, in view of the silence of the contract, the contractor was entitled to all interest earned on retention monies.

In JCT 98, clause 30.5.1 purports to authorise the employer to receive interest on that fund. Because one cannot by contract interfere with trust principles, this clause cannot confer the right on the employer to receive interest on retention monies which are trust funds. There is at present no direct authority to support this proposition, but it is suggested that it is a necessary corollary of the two cases already cited.

Not all retention monies are trust monies. For example, the JCT Minor Works contract does not expressly create a trust in favour of the contractor.

5.10   *Interim certificates: other sums not subject to retention*

In addition to the amounts already specified above **[5.4 and 5.5]**, all of which are subject to the deduction of retention monies, the contract provides in clause 30.2.2 for other sums to be included which are not subject to such deductions.

5.10.1   Under clause 3

Clause 3 provides that where the contract states that an amount is to be added to or deducted from the contract sum, the amount must be included in the next interim certificate after partial or complete ascertainment. The amounts referred to are:

- Fees, rates or taxes (other than VAT): clause 6.2 **[4.14]**
- Architect's instructions concerning opening up if the work is found to be in accordance with the contract: clause 8.3 **[2.5]**
- Payments in respect of patents: clause 9.2 **[3.10]**
- Payments made by the contractor to take out and maintain insurance which it is stated in the appendix may be required and for which the architect has issued an instruction under clause 21.2.1: clause 21.2.3 **[8.6]**.

5.10.2   Loss and/or expense

- Ascertained under clause 26 for matters materially affecting the progress of the work **[10.3]**
- Ascertained under clause 34.3 caused by discovery of antiquities **[10.17]**.

5.10.3   Restoration, replacement or repair and disposal of debris

- Which under clause 22B.3.5 (employer's insurance of new Works) is treated as a variation
- Which under clause 22C.4.4.2 (employer's insurance of work to existing structures) is treated as a variation.

5.10.4   In respect of nominated sub-contractors

- Amounts in respect of their early final payment under clause 35.17

- Amounts in respect of payment of amounts referred to in clause 4.17.2 of NSC/C.

5.10.4   Fluctuations

- Amounts payable under clauses 38 (contribution, levy and tax) or 39 (labour and materials cost and tax) if applicable, but not under clause 40 (use of price adjustment formula). Amounts payable under clause 40 are covered by clause 30.2.1.1 which makes such payment subject to deduction of retention **[5.11]**. The matter, erratically, is also dealt with in the formula clause itself in clauses 40.1.1.1, 40.1.3 and 40.2.

5.11   *Interim certificates: deductions*

Clause 30.2 expressly provides for the amount of any reimbursement of advance payment under clause 30.1.1.6 **[5.20]**, which is due under the terms set out in the appendix, to be excluded from the gross valuation.

The architect is under a duty, in accordance with clause 30.3, to exclude from his certificate the following monies due to the employer which are not subject to retention:

- Amounts deductible under clause 7 which refers to an 'appropriate deduction' being made from the contract sum if the employer consents to the architect instructing that setting out errors are not to be amended
- Amounts deductible under clause 8.4.2 which refers to an 'appropriate deduction' being made from the contract sum if work, materials or goods are not in accordance with the contract and the architect, with the agreement of the employer, allows them to remain
- Amounts deductible under clauses 17.2 and 17.3 which refers to an 'appropriate deduction' being made from the contract sum if the employer consents to the architect instructing that defects are not to be made good
- Amounts allowable by the contractor to the employer under fluctuations clauses 38 or 39 if applicable
- Amounts in respect of appropriate deductions as referred to in NSC/C clause 4.17.3.

The architect is not entitled to certify the value of work which is not in accordance with the contract and, it appears to follow from the principles of *Baston* v. *Butter* (1806), he can abate the sum awarded in respect of sub-standard work which is of less value than if it complied with the contract. As was said in a case which illustrates the distinction between abatement of value and damages for breach, *Mondel* v. *Steel* (1841):

> 'The defendant is permitted to show that the chattel, by reason of non-compliance with the warranty in the one case, and the work in consequence of the non-performance of the contract in the other, were diminished in value: *Kist* v. *Atkinson* (1809); *Thornton* v. *Place* (1832).'

It is clear that interim certificates are each simply one in a series of continuing and cumulative valuations, and if it is discovered that defective work (perhaps not then recognised as such) has been included in an earlier certificate, the architect has a duty to exclude that work from the valuation in the next certificate until, of course, it has been satisfactorily rectified.

5.12   Interim certificate procedure

Clause 30.1.1.1 requires the architect to issue interim certificates which under clause 30.1.3 are to be issued at the intervals specified in the appendix until the certificate following the issue of the certificate of practical completion. The contract provides in clause 17.1 that practical completion will be certified forthwith after the criteria have been satisfied **[12.1]**. However, clause 30.1.3 refers to the factual issue of the certificate of practical completion, '... during which the certificate of Practical Completion *is* issued...'(author's emphasis). Therefore, if for any reason the issue of the certificate is delayed, the effect will be to delay the cessation of regular interim certificates. After this period the architect is only required to issue certificates if further sums are ascertained as payable from the employer to the contractor and after the issue of the certificate of making good defects (to release the second half of the retention) or, if unusually there are no defects, after the end of the defects liability period. There is an overall proviso that the architect cannot be required to issue an interim certificate within one month of the last certificate.

The final date for payment of each interim certificate is 14 days from the *date of issue* of the certificate. The contract does not specify what is meant by that, but it would appear that the operative date is not the one placed on the certificate by the architect, but the actual date on which he despatches it: *London Borough of Camden* v. *Thomas McInerney & Sons Ltd* (1986). Proving the date of despatch may not be easy unless it can be shown that the architect invariably despatched the certificate on the date stated on it.

When the architect considers it necessary to decide how much is due to the contractor, the quantity surveyor must prepare interim valuations (clause 30.1.2.1). If the formula rules for fluctuations are adopted, clause 40 provides that clause 30.1.2.1 is deemed amended to the effect that valuations are made before each certificate. Indeed it is rare for the architect not to require a quantity surveyor's valuation except on the smallest projects. Usually, a routine of application by the contractor (not a contractual requirement, but see [5.23]) and valuation by the quantity surveyor is sensibly established. Clause 30.2 requires that the valuation is to include the period not more than 7 days before the date of the interim certificate. This date is not described as the date of issue and, therefore, refers to the date put on the certificate by the architect. Nothing in the contract states that the contractor must be provided with a copy of the quantity surveyor's valuation unless clause 30.1.2.2 applies.

The amount certified in each interim certificate is to be the gross valuation set out in clause 30.2 from which the retention and previous amounts stated as due have been deducted.

All too often, it seems that architects simply take the valuation produced by the quantity surveyor and copy the figures directly onto to the certificate before signing and issuing it. That is a dangerous procedure. The architect is responsible for the contents of a certificate *(Sutcliffe* v. *Thackrah* (1974)) and he is entitled to certify a different amount to the quantity surveyor's valuation if he believes that this is an appropriate course of action: *R.B. Burden* v. *Swansea Corporation* (1957). Although an architect is obviously entitled to rely on the quantity surveyor's valuation (*R.B. Burden Ltd* v. *Swansea Corporation* (1957)), it is suggested that he owes a duty to the employer to satisfy himself in a general way that the valuation is accurate or perhaps it is better expressed as satisfying himself that it is not seriously inaccurate. In order to accomplish that, he should request the quantity surveyor to furnish him with a brief breakdown accompanying each valuation so that he can see if anything is obviously wrong. Clearly, the architect is not obliged to revalue the work, even if he had the requisite skill and experience.

5.13 *Employer's right to deduct from certified sums*

The employer is entitled by the contract to deduct from sums certified by the architect certain specific and liquidated sums. The most important of these are liquidated damages in the circumstances set out in clause 24.2. Other sums for which express authority is contained in the contract are shown in Table 5.2.

Table 5.2
Sums which the employer may deduct from certified sums

- All costs incurred in connection with the employment of other persons where a contractor fails to comply with architect's instructions: clause 4.1.2
- Insurance premiums paid because of the contractor's default: Clause 21.1.3
- Insurance premiums paid because of the contractor's default: Clause 22A.2
- In any sums due to the employer after determination of the contractor's employment: clause 27.6.6 and 27.7.1
- Amounts under clause 31 (Construction Industry Scheme)
- Amounts paid directly to nominated sub-contractors: clause 35.13.5.2
- The cost of rectifying defects caused by the original nominated sub-contractor who has been paid in full and who refuses to make good: 35.18.1
- The extra cost of a new nominated sub-contractor where the original nominated sub-contractor has validly determined his employment
- Amounts under the VAT agreement.

These provisions leave open the question of whether the employer has a right to refuse payment of certified sums on the ground that he has an equitable right of set-off in respect of a claim to abate the sum due, or counterclaim for unliquidated or quantified damages. JCT 98 does not expressly deal with such claims except in an oblique way via the notice provisions incorporated in accordance with the Housing Grants, Construction and Regeneration Act 1996.

In this as in all other questions of law, the history of the matter is of the utmost importance. There never was a 'common law' right of set-off, for the simple reason that until the start of the nineteenth century contractual promises were enforceable only through the

action of *assumpsit*, which meant that there had to be a separate
action on the promises of each party. This situation was modified by
the Statute of Set-off 1728 which was intended to last for only five
years. It became known informally as the 'Insolvent Debtors Act',
because it created a right of set-off in those circumstances. This is
now dealt with only under the Bankruptcy and Insolvency Acts.

There was another Statute of Set-off in 1734 which lasted until
1879. It is unnecessary to go into the details of all its provisions, but
it applied only where the plaintiff and defendant were mutually
indebted, i.e. each had a claim for a liquidated sum immediately
due: *Crawford* v. *Stirling* (1802); *Morley* v. *Inglis* (1837). Those debts
had to exist when the plaintiff sued, when the defence was filed and
when the trial took place: *Evans* v. *Prosser* (1789). Under the statute,
therefore, there could be no set-off of a claim for liquidated damages, still less for unliquidated damages, which was the nature of
the counterclaim in *Dawnays* v. *Minter* (1971).

By the start of the nineteenth century, the common law would
allow the purchaser of goods an abatement of price for breach of
warranty as to quality: *Kist* v. *Atkinson* (1809); *Thornton* v. *Place*
(1832). But again, there was a firm rule that a counterclaim for
damages of any kind could not be set-off: *Mondel* v. *Steel* (1841).

Equitable set-off was such as was allowed by the Court of
Chancery before its abolition under the Judicature Act of 1873. It
could be a counterclaim for unliquidated damages, but whatever its
nature it was only allowed in circumstances where the court of
Chancery would have granted an injunction to restrain a plaintiff
from enforcing a judgment he had obtained at common law against
the defendant. That is, the nature of the defendant's counterclaim
had to vitiate entirely the judgment that the plaintiff had obtained
or could obtain from the common law courts. The usual example
was a debt incurred by the defendant to the plaintiff as a result of
fraud by the plaintiff.

Equitable set-off must 'directly impeach the right to payment of
the debt', it was said. The Judicature Act 1873 preserved this right to
equitable set-off by section 24 but it passed into complete disuse for
nearly 70 years, until it was resurrected in 1948 by the case of
Morgan & Sons v. *S. Martin Johnson & Co.*

As Master Bickford-Smith commented in a lecture on the
subject:

'When resurrected, the doctrine gave rise to considerable difficulties, both theoretical and practical ... nobody could remember
what the old Court of Chancery did.'

Equitable set-off disappeared again for nearly ten years, until it was once again revived in the case of *Hanak* v. *Green* (1958), which was actually about the costs of building litigation. No reference was made to what had been the practice of the old court of Chancery, and 'equitable set-off' passed into the mythology of English law. The totally irrational belief of many lawyers is that if an employer (or a contractor) puts forward a counterclaim for damages, however spurious it may be, against a certified sum, he is entitled not to have summary judgment given against him.

It is sometimes forgotten that in *Dawnays* v. *Minter*, application was made to the House of Lords for leave to appeal. Although it was on the standard contract, the Green Form, leave was refused. Leading counsel for the applicants conceded that only such sums as the Green Form specifically authorised could be deducted from certified sums. The attitude of the House of Lords was that the decision in *Dawneys* was plainly right. Master Bickford-Smith in his lecture said:

> 'It is undoubtedly the commercial intention of the parties that the building owner should pay the contractor as progress payments the amounts certified on interim certificates by the architect, with no deduction save those expressly authorised by the contract.'

5.14   *Other certificates by architect*

The contract provides for numerous other certificates to be issued by the architect and these are dealt with specifically in the appropriate sections:

Clause 17.1	practical completion **[12.1]**
Clause 17.4	completion of making good defects **[12.4]**
Clause 22A.4.4	payment of insurance monies **[8.8]**
Clause 24.1	failure to complete on the due date **[7.5]**
Clause 27.6.4.2	expenses incurred on determination by the employer **[9.8]**
Clause 35.15.1	failure to complete on the due date by nominated sub-contractors **[4.10]**
Clause 35.16	practical completion of the nominated sub-contract **[4.10]**

5.15   *Validity of the final certificate*

Clause 30.8 requires the architect to issue the final certificate in accordance with a precise timetable. The contents of the final certificate are set out in clause 30.6.2 **[12.7]**.

The final certificate is to be issued within two months of whichever of three specified events is the latest, namely:

- The end of the defects liability period **[12.4]**
- The date of issue of the certificate of completion of making good defects **[12.4]**
- The date on which the architect sent the final account (as set out in clauses 30.6.1.2.1 and 30.6.1.2.2) to the contractor.

Curiously, it is the date *of issue* of the certificate of making good defects which is the operative date here and not the date *named in the certificate* which clause 17.4 states is the date 'for all the purposes of the contract'. Since nothing in clause 17.4 precisely states how promptly the certificate of completion of making good defects must be issued, there may be a substantial difference between the *date of issue* and the date *named*.

A final certificate issued outside the period prescribed may be void: *London Borough of Merton* v. *Lowe and Pickford* (1981). In this case, the architects argued that the final certificate which they issued was invalid, because it was issued about five years late. The reason, for what at first sight appears to be strange behaviour, was that the architects were being sued by the employer on the grounds that the issue of the final certificate had barred them from recovering damages from the contractor for defects which had appeared in the roof. They were also sued in respect of 'design and/or supervision'. The contract was JCT 63 and the final certificate clause was 30(7). Judge Stabb said:

> 'The issue of a proper final certificate in accordance with the terms of clause 30 of the contract would have served as a good defence to any subsequent proceedings that the plaintiffs might have brought against the contractors for bad workmanship on the part of the sub-contractors for which the main contractors would have been liable.
>
> It was contended that the final certificate was invalid and a nullity because it was given out of time ... by reference to the Appendix, it is quite clear that the final certificate should have been given at the latest in 1969.'

It should be noted that the first paragraph of this extract does not apply to JCT 80 after the issue of JCT Amendment 15, or to JCT 98.

The judge also dealt with the contention that the certificate was invalid, because it did not include certain sums which (clause 30.6.2 in JCT 98) stated should be included. He decided that the final certificate was a nullity on both counts. However, the architects were not entitled to set that up as a defence, because to do so would have been to rely upon their own default. In short, they were estopped by their own conduct from relying upon the nullity of the final certificate. In the Court of Appeal decision in *Tameside Metropolitan Borough Council* v. *Barlows Securities Group Services Ltd* (2001), a final certificate was not issued. The contractor subsequently tried to argue that the employer was estopped from relying on the absence of a final certificate, because there had been a meeting between respective quantity surveyors at which there was an agreement or understanding on the matter. The court held that there was no clear evidence that the employer had made any representations that it would not enforce its rights to sue for breach of contract in respect of any defects.

A final certificate may also be void if the architect issuing it is so under the influence of the employer as to cease to be impartial: *Hickman* v. *Roberts* (1913). There, the architect delayed the issue of the final certificate beyond the appointed time and wrote to the contractor advising him to see the employers, because in the face of their instruction to him, he could not issue the certificate whatever his private opinion.

The issue of the final certificate is the last certificate, and probably the last action, which the architect is empowered to take under the contract. Once it is issued, he can do nothing further. He is termed *functus officio* – 'out of office': *Fairweather* v. *Asden* (1979). Similar conclusions were reached in *R.M. Douglas* v. *CED* (1985) and *A. Bell and Son (Paddington) Ltd* v. *CBF Residential Care and Housing Association* (1990).

5.16   The extent to which the final certificate is conclusive

There is a great deal of mythology about the final certificate. The significance of the final certificate has swung violently one way and then the other since the demise of JCT 63.

At one time, when the architect issued the final certificate it was a statement that the whole of the works were complete in all respects in accordance with the contract. That was certainly the position under some early editions of the JCT 1963 form of contract. In later editions of JCT 63 and under JCT 80 the position was substantially modified or so the draftsman thought.

Some forms of contract made the issue of the final certificate conclusive about certain things. Other forms did not state that it was conclusive about anything, not even the amount finally due. The final certificate under the JCT Agreement for Minor Building Works (MW 98) was, and is, an example of the latter category and, therefore, these comments relating to the conclusivity of the final certificate do not apply to MW 98.

When a final certificate is said to be 'conclusive' it means that if neither party has entered into adjudication, litigation or arbitration before the issue of the certificate nor so enters within a stipulated period (28 days under JCT 98) after its issue, the certificate is conclusive (i.e. unchallengeable) evidence in any such proceedings in regard to the stipulated matters. Thus, if a final certificate is said to be conclusive in regard to the amount of the final sum certified, it will not prevent an aggrieved party from seeking satisfaction by way of arbitration if the sum is considered to be wrong: *P. & M. Kaye Ltd* v. *Hosier & Dickinson* (1972). However, the other party has simply to produce the final certificate for the matter to be at an end. Under JCT 80, final certificates were conclusive in respect of the following:

- *Where the quality of materials and standards of workmanship are to be to the reasonable satisfaction of the architect, the architect is so satisfied*: This clause referred to clause 2.1 stating the contractor's obligations. Failure to realise what this meant gave rise to many misconceptions. Part of the contractor's obligations was to ensure that if the architect had stated that certain things were to be to the architect's satisfaction, such things were to his or her satisfaction. This may have been done by stating that specified items must be 'approved' or 'to the architect's satisfaction' or some other form of words to the same effect. When the final certificate was issued, it was conclusive evidence that the architect was satisfied with any matters which were so specified whether or not the architect had in fact specifically expressed approval or even looked at the item in question. It will readily be appreciated that to insert some such phrase as 'All workmanship and material unless otherwise stated, must be to the architect's satisfaction' was opening the door to the blanket conclusivity of the final certificate again. It was the business of the architect's satisfaction and the willingness of the courts to give a very wide interpretation to matters which ought to be to the architect's satisfaction which gave rise to problems.

- *All the provisions of the contract requiring adjustment of the contract sum have been complied with:* the final certificate was conclusive evidence that all necessary adjustments had been properly carried out. Claims by the contractor after the appropriate period had elapsed from issue of the certificate, that the figures were wrong would be fruitless. The only exceptions were if there had been accidental inclusion or omission of work or materials, fraud or if there is an obvious arithmetical error. This sub-clause is unchanged.

- *All due extensions of time have been given:* This was to prevent the contractor raising the question after the final certificate when the employer may have deducted liquidated damages and all financial matters appeared to have been settled. This sub-clause is also unchanged.

- *That reimbursement of loss and/or expense is in final settlement of all contractor's claims in respect of clause 26 matters whether the claims are for breach of contract, duty of care, statutory duty or otherwise:* This is a very widely drawn clause intended principally, like the previous clause, to ensure that the final certificate really does spell the end of the financial road. It should be noted, however, that the conclusivity is effective only in respect of the clause 26 matters. It will not operate to prevent the contractor from making claims in regard to breaches of contract in respect of occurrences outside the clause 26 matters. This sub-clause still applies unchanged.

The effect of the issue of the final certificate, especially in regard to the architect's satisfaction with workmanship and materials, was considered by the Court of Appeal: *Crown Estates Commissioners* v. *John Mowlem & Co* (1994). Much to the concern of architects, the court decided that the final certificate under JCT 80 was conclusive that the architect was satisfied with the quality and standards of *all* materials, goods and workmanship. The consequence of that was that it was virtually impossible for the employer to take action against the contractor for latent defects after the issue of the final certificate.

When the Court of Appeal considered the effect of the final certificate and came to its decision, it was not making new law. It was telling everyone what the terms of the contract meant even though until that moment perhaps no one (including the Court of Appeal) had realised it. What the parties intend to do when they enter into a contract is, of course, important, but only in so far as they give effect to their intentions by the written terms they agree in the contract. In turn, the court can only interpret their intentions by looking at the contract terms. Evidence as to their intentions outside a written contract is normally inadmissible. The court's interpretation of the contract term was very much in the contractor's favour, but that is what the parties agreed in law when they signed the contract.

The terms of the contract required the architect to issue a final certificate within a specific timescale. If the architect did not so issue he, and through him the employer, was in breach. The architect's position was straightforward if he or she had been engaged on the usual (SFA/99) terms of engagement or similar. During the progress of the Works the architect must carry out inspections with reasonable skill and care. There is a duty to issue the final certificate in accordance with the building contract. If subsequently, a latent defect was discovered, the employer may have been unable to recover the cost of remedial work from the contractor. The employer might then have turned attention to the architect. If the employer was to have been successful in recovering the loss from the architect in negligence, it would have to be shown that the architect failed to carry out administrative duties, including inspection, with reasonable skill and care. That would not have been very easy, but perhaps easier than most architects would have wished.

It was common, in former times, for architects to be so concerned about the conclusiveness of the final certificate that they often neglected to issue a final certificate at all, leaving a minute sum of money outstanding in the knowledge that the contractor would not seek arbitration in respect of such a small amount. By this method it was hoped that the matters otherwise made conclusive by the final certificate would be left open and the employer would not be precluded from obtaining redress from the contractor if any latent defects appeared. Of course, it had also to be borne in mind that, if successful, such a ploy would effectively deprive the employer of the conclusive benefit of the other three matters. The courts have put an end to any likelihood that an employer could proceed against the contractor if the final certificate was not issued at the proper time. The court's view appears to be that if the failure to issue was a breach of contract, the employer cannot take advantage of that breach: *Matthew Hall Ortech Ltd* v. *Tarmac Roadstone Ltd* (1997). However, the situation is likely to be different if the contractor has not bothered to request a final certificate: *Tameside* v. *Barlows Securities* (2001).

Following the *Crown Estates* case, the JCT issued amendments to each of the affected forms of contract which were intended to remove the effect of the Court of Appeal decision by rewording the sub-clauses relating to the conclusivity of the architect's satisfaction. Essentially, therefore, the position was restored that the final certificate was conclusive about the architect's satisfaction only if the architect had specifically stated in the bills of quantities or specification that some item of goods, materials or workmanship was to be to his or her satisfaction or approval.

The introduction of the adjudication procedure has complicated the calculation of the date on which the final certificate can be said to be conclusive under JCT 98. That is because the adjudication decision has only a temporary binding quality. If proceedings in adjudication, arbitration or litigation have been started before the issue of the final certificate, it will not be conclusive until the earliest of either:

- The proceedings have finished when the final certificate is subject to the decision in each case or, perhaps surprisingly, any settlement of the proceedings

or:

- 12 months after the issue of proceedings, if neither party has taken any step in the proceedings. The final certificate is then subject to partial settlement terms (if any).

The position is different if such proceedings were not commenced until after the final certificate was issued, but before the expiry of 28 days from the date of issue. In that case, the final certificate is conclusive except in relation to the subject of the proceedings.

If the adjudicator gives his decision after the issue of the final certificate and either party wishes the subject matter to be referred to arbitration or, if applicable, dealt with in litigation, they have 28 days from the date on which the adjudicator gave his decision to start these proceedings. It, therefore, appears that if the contractor decides to refer a matter to adjudication after the issue of the final certificate, but within the 28 day period, the final certificate will become conclusive within 28 days except for the subject matter of the referral. If the adjudicator's decision is given, say, 20 days after the final certificate would otherwise become conclusive, the contractor has a further 28 days in which to submit the matter referred to the adjudicator, but no other matter, to either arbitration or litigation as applicable.

5.17   *Fluctuation clauses*

The provisions for fluctuations are referred to in clause 37 and contained in three separate clauses: 38, 39 and 40. Clause 38 is limited to statutory and other charges and alterations. Clause 39 is the full fluctuation clause. Clause 40 is the formula method based on monthly indices.

Unless clause 39 or 40 are specified in the appendix, clause 37.2 states that clause 38 will apply. This clause allows fluctuations relating to 'contribution, levy and tax' payable by the contractor in his capacity as an employer. In its original form, it was designed to allow the contractor relief from impositions such as Selective Employment Tax.

5.18   *Fluctuations frozen on due completion dates*

Whichever method of calculating fluctuations is chosen, the contractor's right to receive payment for subsequent increases ceases on the date on which completion should have taken place: that is, the original contract date or such later date as has been fixed by the architect as the result of extensions of time given under clause 25. The relevant clauses are 38.4.7, 39.5.7 and 40.7.1.

In each case, the right to freeze fluctuations is lost if clause 25 is amended or if the architect does not respond to each written notification of delay received from the contractor (see clauses 38.4.8, 39.5.8 and 40.7.2). Quantity surveyors are apt to amend clause 25 to delete the relevant event dealing with the availability of labour or materials. So far as fluctuations are concerned, that approach is disastrous. Architects cannot simply put consideration of the contractor's notice on hold if an extension of time is not thought to be due. The architect must respond, either positively or negatively. These clauses can, of course, be deleted so as to preserve the right to freeze fluctuation even where amendments are made to clause 25 or the architect fails to respond. Certainly, they cannot be ignored although, worryingly, anecdotal evidence suggests that they are not widely known.

These provisions in effect impose a monetary penalty.

5.19   *Productivity and bonus payments*

In *William Sindall Ltd* v. *North West Thames Regional Health Authority* (1977) increased costs by a bonus scheme, recommended by the

National Joint Council for the Building Industry (NJCBI), but not obligatory, were held by the House of Lords not to be an eligible fluctuation in wages for the purposes of JCT 63 clause 31D. JCT 98 alters that ruling by providing that the prices in the contract bills are based upon the rates of wages and other emoluments payable by the contractor in accordance with any incentive scheme, productivity agreement under the provisions of Rule 1.16 or any successor of the Rules of the Construction Industry Joint Council or some other wage-fixing body (clause 39.1.1.4). The effect of JCT 98 clause 39.1.2 appears to be that any increases in such productivity or bonus payments will henceforth be recoverable by the contractor.

The wording of clauses 39.1.1.4 and 39.1.1.5 is wide enough to cover many different incentive schemes and may, therefore, constitute a trap for the employer and his architect, who may not investigate carefully each separate tender in this respect.

5.20 Advance payment

All the JCT contracts, other than MW 98 and the local authority versions, make provision for advance payment of the contractor if the employer so desires. Why local authorities are excluded is a mystery, because local authority departments have a fairly rigid spending allocation and departments are frequently in the position of wishing to spend the remainder of an allocation before the end of the financial year. Advance payment appears to be an ideal way of achieving this.

In JCT 98, clause 30.1.1.6 deals with the procedure. The basic idea is that, if the contractor agrees, the employer may make a payment to the contractor at the beginning of the contract Works. The contractor repays the amount in instalments spread over an agreed period.

The appendix must state that clause 30.1.1.6 applies and it must also state the amount and the date on which the payment is to be paid by the employer. The terms of reimbursement are also to be stated.

The contract makes provision for the employer to require a bond before making payment. It is difficult to see why the employer would not require a bond and the procedure would have been simplified if the bond requirement had been stated to be automatic. If a bond is required, the appendix must record it and the terms of the bond must be annexed to the appendix. The contract has bound into it a sample of the bond agreed between the British Bankers'

Association and the JCT. If the employer wishes to use a bond on other terms, the seventh recital makes clear that the contractor must have been given copies of the terms, presumably at tender stage. If the employer is satisfied with the bond attached to the contract, the seventh recital should be struck out.

The surety is subject to the employer's approval. If the contractor defaults on the agreed reimbursement schedule, there is a sample notice of demand which the employer may use to obtain payment from the surety. Surprisingly, the reimbursement to the employer is to be effected by means of the architect's certificate. Clause 30.2 lists the part due of the advance payment as one of the items which the architect must deduct from the gross valuation. Therefore, in the first instance, the payment which the contractor is to make is reimbursed to the employer under the certification procedure. It is likely that the reimbursement terms will provide for a series of monthly repayments, which may of course include provision for interest. Under this system, difficulties will arise if a monthly gross valuation before deduction of the repayment is less than the repayment. The architect will be faced with issuing a negative certificate. It would have been simpler and made good practical sense if the repayments had been achieved by a withholding notice from the employer. That would have kept the employer's loan quite separate from the other financial considerations of the contract.

5.21   *Withholding notices*

Consequent upon the Housing Grants, Construction and Regeneration Act 1996, all construction contracts must contain provisions for the giving of notices in two important situations: the amount to be paid and the amount, if any, to be withheld. JCT 98 is no exception. The provisions are set out in clauses 30.1.1.3, 30.1.1.4 and 30.1.1.5.

Under clause 30.1.1.3, the employer must issue a written notice to the contractor stating the amount of payment he proposes to make and the basis on which it is calculated. The notice must be sent no later than five days after the date of issue of an interim certificate. The date of issue of a certificate is not the same as the date of the certificate. The date of the certificate is the date *on* the certificate. The date of issue is the date the architect does a formal act which can be interpreted as issuing (*London Borough of Camden* v. *Thomas McInerney & Sons Ltd* (1986)), for example, putting it in the post.

At first sight, the wording of the clause appears to give the employer the option to state a sum less than the certified amount provided he demonstrates how it is calculated. That does not reflect what the Act says. The relevant part of the Act is section 110(2):

> '(2) Every construction contract shall provide for the giving of notice by a party not later than five days after the date on which a payment becomes due from him under the contract, or would have become due if –
> (a) the other party had carried out his obligations under the contract, and
> (b) no set-off or abatement was permitted by reference to any sum claimed to be due under one or more other contracts, specifying the amount (if any) of the payment made or proposed to be made, and the basis on which the amount was calculated.'

Clearly, the Act does not especially envisage that a certificate will have been issued under the terms of a contract in which employer and contractor will have agreed that the certificate will state the amount due. Although some commentators have speculated that the wording of clause 30.1.1.3 gives the employer power to state a lesser sum than the sum certified, there appears to be no procedure for so doing. Indeed, it is suggested that the employer's power is restricted to stating the amount certified as the amount to be paid in the written notice.

If the employer for any reason wishes to pay a lesser amount, he must observe the procedure in clause 30.1.1.4 which requires him to give a written notice not later than five days before the final date for payment. The final date for payment is 14 days after the date of issue (not the date of the certificate) of an interim certificate. The notice must specify the amount withheld and the ground for withholding. If there is more than one ground, the amount in respect of each ground must be expressly set out together with detailed reasons. For example, it is not sufficient for the employer simply to state: 'I do not agree that you are entitled to the amount certified, because there is a lot of defective work and I am going to deduct £5000'. It is sometimes thought that, either under clause 30.1.1.3 or clause 30.1.1.4, the employer may withhold payment on the basis that he is abating the price. The argument goes that he is not deducting money, he is simply not paying as much as certified, because the work is not worth the certified amount. The distinction may be more theoretical than actual, but in any event, it appears that a written notice under clause 30.1.1.4 must be given whether set-off or abatement is the reason for withholding. Judge Bowsher

QC said in *Whiteways Contractors (Sussex) Ltd* v. *Impresa Casteli Construction UK Ltd* (2000):

> 'It is common for a party to a building contract to make deductions from sums claimed on the Final Account (or on earlier interim applications) on account of overpayments on previous applications and it makes no difference whether those deductions are by way of set-off or abatement. The scheme of the Housing Grants, Construction and Regeneration Act 1996 is to provide that, for the temporary purposes of the Act, notice of such deductions is to be made in a manner complying with the requirements of the Act. In making that requirement, the Act makes no distinction between set-offs and abatements. I see no reason why it should have done so, and I am not tempted to try to strain the language of the Act to find some fine distinction between its applicability to abatements as opposed to set-offs.'

This appears to be an exceptionally clear statement of the position. It is difficult to see why some commentators cite this case as authority that abatement does not require a withholding notice.

The same principle holds good for the provisions in JCT 98. Clause 30.1.1.4 largely reflects what the Act says in section 111. However, the second paragraph of section 111(1) states that a written notice given in accordance with section 110(2) may suffice as a notice of intention to withhold payment if it complies with the requirements of section 111. Clause 30.1.1.4 does not say that, but the Act would take precedence. Therefore, if a written notice under clause 30.1.1.3 of the contract also complied with the requirements of clause 30.1.1.4, there would seem to be no reason why the one (earlier) notice could not take the place of both.

Clause 30.1.1.5, however, introduces more uncertainty. It states that if the employer fails to give 'any written notice' under clause 30.1.1.3 'and/or' clause 30.1.1.4 the employer must pay the certified amount. The precise meaning of 'and/or' in these circumstances is not clear. To be sure, the employer should issue both notices at the relevant times, otherwise under the terms of the contract, as compared to the Act, the employer will have difficulty in avoiding payment of the certified sum in full. Although the Act may allow service of the earlier notice (under section 110) to serve the purposes of both notices, and the employer may satisfy the requirements of the Act by so doing, the particular and more onerous terms of the contract may require payment. There is nothing to prevent the service of both notices at the same time, however, provided that is the earlier time.

The same notice provisions apply to the final certificate. In that instance, clause 30.8.2 deals with the initial written notice, but instead of referring to the amount certified, it refers to 'any balance stated as due to the Contractor'. This is presumably to acknowledge the situation in which the balance is shown to be due from the contractor to the employer which is expressly envisaged by clause 30.8.1.

The final date for payment of any balance due to the contractor is stated by clause 30.8.3 to be 28 days from the *date of issue* of the final certificate and the comments earlier in this section on the topic of 'date of issue' of the certificate apply to the final certificate also. This clause also reflects the contents of clause 30.1.1.4 in respect of the withholding notice which must be issued no later than five days before the final date for payment.

5.22   *Interest on late payments*

Clause 30.1.1.1 refers to interest on late payment of interim certificates and clause 30.8.5 refers to interest on late payment of any balance in the final certificate. Essentially, the terms are the same, but there is a particular difference which bears consideration.

The general position at common law is that there is no automatic right to interest if payment is late. Certain requirements must be met. There must be either an express or implied term to that effect, or the interest must be payable either as part of damages or awarded by the court. The Late Payment of Commercial Debts (Interest) Act 1998 has modified that position. It came into force on 2 November 1998 and it becomes fully effective after 2 November 2002. The Late Payment of Commercial Debts (Rate of Interest) Order 1998 fixed the current rate of interest as 8% above Bank of England Base Rate. The Act provides, by section 8(2), that a contractual remedy can displace the Act if the remedy is 'substantial'. JCT 98 provides an interest rate of 5% above Bank of England Base Rate. It is thought that this rate would be considered 'substantial' for the purposes of displacing the remedy in the Act.

Clause 30.1.1.1 refers to interim certificates and provides that if the employer does not properly pay any part of the amount 'due to the Contractor under the Conditions by the final date for payment', the employer must pay simple interest as well as the amount due until all amounts owing are discharged. It is to be noted that although the first part of the clause deals with interim certificates, the part dealing with interest makes no reference to the employer's failure to pay the certified sum, but only his failure to pay an amount due 'under the Conditions'. This must, at the very least, leave open the possibility that the contractor could claim interest on all monies which can objectively be shown to be due under the contract terms even if the architect has failed to certify them.

Clause 30.8.5 expressly refers to the employer's failure to pay the 'said balance' or any part of the balance. It can only refer to the balance shown in the final certificate and, therefore, the possibility of claiming interest without a certificate is not available under the final certificate.

In both clauses, the 5% is to be added to the Bank of England Base Rate which is current when the employer's payment becomes overdue. There is no provision for the rate to be adjusted in line with adjustments to the Base Rate and, therefore, it seems that once the interest rate has been fixed by reference to the Base Rate current on the day after the final date for payment, the rate stays the same until the employer has paid all monies owing. It will be a matter of luck whether the Base Rate is especially high or low on the relevant date.

Although it is not necessary to do so, the clauses make clear that payment of interest by the employer will not be construed as a waiver of the contractor's right: to the principal sum owing, to the time of payment, to the sum being paid under the contract terms, to suspension of performance of obligations under clause 30.1.4 **[9.15]** and to determine his employment under clause 28.2.1.1. Put simply, the contractual right to interest is additional to, and not instead of, the contractor's other rights under the contract. No doubt it was thought appropriate to set out the position to avoid lengthy disputes on the matter during which the employer may use the sum owing as a loan facility – albeit at a high rate of interest.

5.23   Contractor's application

One of many new features of JCT 98 is the right for the contractor to make application for payment under interim certificates. This right is contained in clause 30.1.2.2. Clause 30.1.2.1 refers to the quantity surveyor's duty to make valuations whenever the architect considers them to be necessary **[5.12]**. Clause 30.1.2.2 is stated to be without prejudice to the architect's obligation to issue interim certificates. So it is clear that whether or not the contractor makes application, the architect must certify in accordance with the contract terms the amounts properly due.

It is common practice for contractors to make application for payment in any event, but this clause puts the practice on a formal footing. The contractor need not make application, but if he does, he has until seven days before the date of an interim certificate (not the date of issue of the certificate) to submit his application to the quantity surveyor. The application is not the opportunity for the contractor to engage in flights of fancy or indeed to devise interesting new methods of valuation. The application must be based on the valuation rules in clause 30.2. The contractor must include in the application any applications made by nominated sub-contractors with gross valuations under NSC/C clause 4.17.

Although clause 30.1.2.1 refers to the quantity surveyor's duty to make valuations whenever the architect considers them to be necessary, the effect of the contractor's application is to trigger the quantity surveyor's valuation whether or not the architect has requested it. If the quantity surveyor disagrees with any part of the contractor's application, he must send the contractor a statement setting out the basis of the disagreement in the same amount of detail as was contained in the contractor's application. This is to be done at the same time as the issue of the quantity surveyor's formal valuation to the architect.

Two things flow from this. The first is that it will be in the contractor's interest to produce an application in as much detail as possible, because the response must be in similar detail. The quantity surveyor is not entitled to say simply that he has arrived at a different figure. The contractor, by carefully detailing the application, can ensure that the quantity surveyor has a contractual obligation to provide a similar response. At least, the contractor should know why he is not getting paid. The second thing is that a failure to adequately respond will be a breach of contract on the part of the quantity surveyor. Quantity surveyors are so used to the usual routine of receiving the contractor's application and using it as a basis for the valuation that they, and the contractor, may well forget the new provisions. The contractor's redress for such breach, however, appears to be merely the right to insist on a response from the quantity surveyor. By the time it has been before an adjudicator the matter could be academic.

5.24   *The priced activity schedule*

The second recital records that the contractor has provided the employer with a fully priced copy of the bills of quantities. The second part of the recital goes on to say that the contractor has also provided a priced activity schedule. This part must be deleted if a priced activity schedule is not provided. If it is provided, it must be attached to the appendix. The 'priced activity schedule' is not defined. However, and curiously, the 'Activity Schedule' is defined in clause 1.3 as the schedule of activities attached to the appendix 'with each activity priced'. If that is the definition of the 'Activity Schedule', it is difficult to understand the effect of adding the word 'priced' as is done in the second recital and in clause 30.2.1.1. The definition proceeds to say that the sum of the individual prices is the contract sum, but excluding provisional sums, prime cost sums, contractor's profit and the value of work in approximate quantities included in the contract bills.

The full import of what this book will continue to call the priced activity schedule (for no better reason than that the second recital and clause 30.2.1.1 refer to it in similar fashion) is set out in clause 30.2.1.1. If the appropriate entry in the appendix is completed to indicate that a priced activity schedule is attached, it is to be used for the interim valuation of the activities listed. That is to be accomplished by proportioning the value of the work properly executed. For example, if the priced activity for a heating system is £120,000 and the quantity surveyor decides that 75% is properly completed, the amount added to the valuation in respect of the heating system would be £120,000 × 75% = £90,000. This will be a fairly rough and ready system of valuation, but perhaps in practice it will be no more approximate than the supposedly precise system based upon priced bills of quantities.

It will be noted that priced bills are not superseded by the priced activity schedule. The valuation of variations must be dealt with under clause 13 which refers exclusively to the bills of quantities [6.4].

CHAPTER SIX

VARIATIONS AND THEIR VALUATION

6.1 Does extra work always involve payment?

Extra work in a broad sense must mean work which is extra or additional to the Works which the contractor is required to execute in accordance with the contract. This can arise by a straightforward addition to the contract Works, as in an architect's instruction to the contractor to build an additional three courses of brickwork at the top of a wall which is already specified and for which a price is included in the bills of quantities. Alternatively, extra work may arise because the architect has instructed the omission of something and the substitution of something else. For example, the architect may omit a bath and instruct an elaborate shower and cubicle to be installed instead. The net result may be an increase in the cost of the Works. The valuation of variations is dealt with under the very complex clauses 13 and 13A.

JCT 98 contains detailed requirements which must be satisfied before the contract sum can be adjusted [6.2]. The fact that the contractor has carried out work which is additional to what the contract requires does not automatically entitle him to additional payment. On the contrary, if the work is not instructed, the contractor is probably in breach of his obligations. This is a fundamental principle. Were the situation to be different, there would be nothing to prevent the contractor, in need of extra profit, to unexpectedly do some work not included in the contract and claim payment simply on the ground that he had done it. There would be nothing to prevent a contractor supplying gold bath taps instead of the specified chrome taps and claiming the difference in cost. However, identifying whether or not extra work attracts payment may not always be so easy.

The contractor will never be entitled to extra payment if he has to carry out work included in the contract, but which he overlooked or misunderstood when pricing. Whether an item of work was included in the contract Works can be a source of dispute.

6.2   *What is a variation?*

'Variation' is defined in clause 13.1. It is stated to mean two rather different things: first, the alteration or modification of design, quality or quantity of the Works (13.1.1), and second, the imposition or change by the employer of obligations or restrictions (13.1.2) **[2.6]**. But it is stated to exclude the nomination of a sub-contractor to undertake the supply and fixing of materials or carrying out of work for which measured quantities have been set out in the bills of quantities for the contractor to execute. This is on the simple principle that to take work from the contractor and give it to another is a breach of contract **[6.5]**.

The first part of the definition is enlarged by the provision of what appear to be examples in clauses 13.1.1.1, 13.1.1.2 and 13.1.1.3 prefaced by the word 'including'. Clause 13.1.1.1 is hardly worth including, because it refers to 'addition, omission or substitution' of work which must be part of 'alteration or modification. . .'. A similar criticism can be levied at clause 13.1.1.2 which refers to the alteration of the kind or standard of materials or goods. Only clause 13.1.1.3 contributes a situation which might not at first sight be envisaged (but certainly would on further consideration) by the brief first part of this definition. It refers to removal of work or materials, other than defective work or materials, from site. The first part of the definition is, therefore, extremely broadly drafted and refers to any change of any kind at all in the Works.

The second part is specific. The obligations or restrictions must concern one or more of the following:

- Access or use of any part of the site
- Limitation of working space or hours
- Carrying out the work in a particular order.

This part is more akin to varying the contract terms than the work content. Indeed, it clearly does not vary the Works in any way, but only the way in which they are executed. It should be noted that although the order of the work may be changed, or indeed created if there is no order stated in the bills of quantities, there is no provision to insert any specific dates against the parts of the work. Therefore, although the order may be stipulated, the contractor is under no obligation anywhere in the contract to complete the parts by any particular dates and the architect clearly has no power under clause 13.1.2 to require completion of the parts by any dates other than the date for completion of the whole of the Works as stated in the appendix. The only way to ensure that the contractor has an obligation for completion of individual parts of the Works by specific dates is to use the Sectional Completion Supplement.

Clause 13.2.1 confirms that the architect may issue variation instructions. This is one of his most important powers, sadly overused in many cases. In addition, clauses 13.3.1 and 13.3.2 place a duty on the architect to issue instructions regarding the expenditure of provisional sums whether in the bills of quantities or in a nominated sub-contract.

6.3   *Payment on an oral instruction*

The contract only recognises written instructions (clause 4.3.1) or instructions which have been confirmed in writing by either architect or contractor within the stipulated timescale or confirmed in writing by the architect before the issue of the final certificate (clause 4.3.2) **[2.9]**. The contract makes no express reference to oral instructions neither does there appear to be any method by which the contractor can be paid for complying with them if the architect refuses to use his power to ratify.

The question was considered in *W.S. Harvey (Decorators) Ltd* v. *H.L. Smith Construction Ltd* (1997). The contract specified that instructions must be in writing and that if the contractor issued oral instructions they may be confirmed in writing by the contractor or the works contractor within seven days. If they were not confirmed, they were to have no effect. The contract stated that the works contractor would not be entitled to payment without that written instruction. The judge was quite clear that without confirmation, the instructions were ineffective and the contractor was not to be paid.

However, *Redheugh Construction Ltd* v. *Coyne Contracting Ltd and British Columbia Building Corporation* (1997), a Canadian case, held that if the contract contains a term that all instructions must be in writing and the employer issues an instruction without written confirmation, he may be held to have waived the term so that he cannot assert a breach of contract nor refuse payment. *Ministry of Defence* v. *Scott Wilson Kirkpatrick and Dean and Dyball Construction* (2000) is to much the same effect. The contract stated that oral instructions must be confirmed in writing. However, the court held that it was not necessary for instructions to be in writing in order to be effective.

6.4   *Valuation of variations*

What was, in the original JCT 80, a detailed provision setting out how variation should be valued by the quantity surveyor, has had two lengthy sections added so that there are now three separate systems of valuing variations under JCT 98:

- Alternative A: the contractor's price statement
- Alternative B: valuation by the quantity surveyor
- Clause 13A: the contractor's quotation at the architect's instruction.

The result is a clause of such mind-blowing complexity that it is difficult to comprehend what JCT had in mind when they allowed this monstrous thing to evolve. It is certainly not a simple-to-operate clause readily comprehensible by all parties.

The key to understanding the clause is contained in clause 13.2.3 which sets out the order in which the valuation methods will apply. However, there are difficulties. Nothing anywhere in the two-page clause 13A states that valuation using clause 13A quotation is triggered by a statement to that effect in the instruction. Quite the contrary. Clause 13A states that it only applies to an instruction if the contractor has not disagreed under clause 13.2.3. Clause 13.2.3, however, states that the valuation of a variation must be in accordance with clause 13.4.1.1 'unless the instruction states that the treatment and valuation of the Variation are to be in accordance with clause 13A...'. The position seems to be that, although the architect has no express power to request a clause 13A quotation, without such request clause 13A will not apply. On that basis, it is difficult to see when clause 13A will ever apply. If the instruction, however, contains such a request, the contractor has seven days, unless the parties have agreed another period, within which to disagree in writing that clause 13A should apply. There is no requirement that the contractor's disagreement should give reasons or, indeed, be reasonable and it is doubtful that it would be implied. Once the disagreement has been registered by the contractor, he has neither the right nor the obligation to carry out the variation until the architect issues a further instruction that it is to be carried out and valued under clause 13.4.1.

Clause 13.2.3 goes on to state that the other exception to the application of clause 13.4.1.1 is if clause 13A.8 applies. Clause 13A.8 simply says that if the architect's instruction refers to work which has already been valued under the clause 13A procedures, the quantity surveyor must carry out the valuation 'on a fair and reasonable basis' with reference to the previous clause 13A quotation and must include the direct loss and/or expense resulting from the material affectation of the progress of the Works due to compliance with the instruction. This is a provision which is very similar to the valuation provision in MW 98 clause 3.6.

If the contractor, with the agreement of the architect, has tendered for work which is otherwise to be the subject of a prime cost sum and executed by a nominated sub-contractor and if the tender is accepted, the valuation of the work must be done by reference to the contractor's tender and not to the contract bills. The situation is dealt with by clauses 13.4.2 and 35.2 **[4.7]**.

6.5   *Alternative B*

It is worth looking first at alternative B, being essentially the valuation clause as originally included in JCT 80, albeit now with many amendments. It is still the basic valuation position. The first thing to note is that the valuation is to be made by the quantity surveyor. The valuation rules are contained in clauses 13.5.1 to 13.5.7. The basic rules in clause 13.5.1 are prefaced by the stipulation that they relate to that amount of work which is additional or substituted and either it can be properly valued by measurement or it is such work as has an approximate quantity in the bills of quantities.

The work is to be measured 'in accordance with the same principles as those governing the preparation of the Contract Bills', i.e. normally (unless clause 2.2.2.1 is amended to provide otherwise) in accordance with the Standard Method of Measurement of Building Works, 7th edition, and is to be valued in accordance with the rules set out in clause 13.5.1. So far as variations are concerned, this obviously only applies to variations in the actual work to be carried out as defined in clause 13.1.1, to the extent that the work can reasonably be measured and is not of such a nature that it can only be valued on the basis of 'prime cost', i.e. as daywork. The rules of valuation are set out in clauses 13.5.1.1 to 13.5.1.5 supplemented by clauses 13.5.3, 13.5.3.3, 13.5.5 and 13 5.7. In the basic rules set out in clause 13.5.1 there are three principal factors to be taken into account: the character of the work; the Conditions under which the work is carried out; and the quantity of the work:

(1)   If all three factors are unchanged from an item of work already set out in the contract bills, the price in the contract bills for that item must be used for the valuation of the variation.

(2)   If the character is 'similar' to that of an item of work set out in the contract bills, but the conditions are not 'similar' and/or the quantity is significantly changed from the contract bills, then the price set out in the contract bills against that item must be used as the basis of the valuation of the variation but must be modified so as to make a 'fair allowance' for the changed conditions and/or quantity.

(3)   If all three factors are changed, then the items are to be valued by the quantity surveyor at 'fair rates and prices'.

If the character of the variation is 'similar' to that of an item in the contract bills these rules mean that the quantity surveyor must use the rates set out in the bills for their valuation, and the only modification he can make is in respect of changes in conditions or significant changes in quantity. Only if the character is not 'similar' is the quantity surveyor given complete discretion to make what, in his opinion, is a fair valuation of the work. Therefore, it is essential that the meaning of the phrase 'similar character' should be clearly understood.

In everyday language, 'similar' means 'almost but not precisely the same''or 'identical save for some minor particular'. The words 'similar character' when applied to an individual measured item of work probably mean that the item is identical to an item in the contract bills. If the item is of 'similar character' the only grounds upon which the quantity surveyor can vary the price for the item from that which is set out in the bills is that the conditions are not similar or the quantity has significantly changed, otherwise he must use the price in the bills as it stands. It seems that very little change in description would be needed to render the character of work dissimilar for the purpose of this clause. Then, the rules set out in clauses 13.5.1.1 and 13.5.1.2 cannot be applied and the quantity surveyor is given discretion under clause 13.5.1.3 to value the item at a 'fair' rate or price.

In deciding whether the rates and prices in the bills are to be ignored and a valuation at 'fair rates and prices' substituted, something more than a look at the description or measurement is required. So far as 'conditions' are concerned, it is not necessary to apply the same strict interpretation to the word 'similar'. The 'character' of an item is precisely defined by its description in the bills; the 'conditions' under which it is to be carried out cannot be so precisely defined, and the question of whether the conditions are 'similar' must be judged by vaguer criteria related to the conditions under which the contractor must be deemed to have anticipated that the work in the original bills would be carried out and those under which the varied work actually was carried out. The quantity surveyor is not entitled to take into account the background against which the contract was made: *Wates Construction (South) Ltd* v. *Bredero Fleet Ltd* (1993). The 'conditions' referred to in the valuation rules are the conditions to be derived from the express provisions of the contract bills, the drawings and other documents.

When dealing with quantity, the key factor is not whether it is similar, but whether or not the quantity has been 'significantly changed' by the variation. No firm rules can be laid down. This must be a matter for the objective judgment of the quantity surveyor; a small change in quantity may be significant for some items (particularly where the original quantity was small) but a very large change may not be significant for others.

The rate or price may have been inserted by the contractor in error and perhaps can be conclusively demonstrated to be inaccurate, whether in respect of being too high or too low. It is of no consequence; the rate or price in the bills must be used as the basis and only adjusted to take account of the changed conditions and/or quantity. The contractor has contracted to carry out variations in the work, and the employer has contracted to pay for them on this basis, and neither can avoid the consequences on the grounds that the price in the bills was too high or too low: *Henry Boot* v. *Alstom Combined Cycles* (1999). A contractor will sometimes take a gamble by putting a high rate on an item of which there is a small quantity or a low rate on an item of which there is a large quantity in the expectation that the quantities of the items will be considerably increased or decreased respectively. If the contractor's gamble succeeds, he will make a nice profit. It is not unlawful, but rather part of a contractor's commercial strategy: *Convent Hospital* v. *Eberlin & Partners* (1988).

The simple rule for the valuation of omissions from the contract works is that they are to be valued at the rates set out in the contract bills. However, under clause 13.5.5, if the omissions substantially change the conditions under which other work is executed it must be dealt with accordingly. It is a basic principle that (in the absence of express conditions to the contrary in the contract), once a man has contracted to do a certain quantity of work, he has the right to do it if it is to be done at all; if the contract so provides, the work may be omitted, but only if it is not to be done at all, not in order to give it to someone else. The cases most generally quoted as authority for this proposition are Australian: *Carr* v. *J.A. Berriman Pty Ltd* (1953); *Commissioner for Main Roads* v. *Reed & Stuart Pty Ltd & Another* (1974).

An American case is even closer in comparison to the conditions of the JCT contracts: *Gallagher* v. *Hirsch* (1899). There, as in the JCT forms, the contract provided for the omission of work without vitiating the contract and provided that such omissions should be valued and deducted from the contract sum. The American appeal court held that the word 'omission' meant only work not to be done at all, not work to be taken from the contractor under the contract and given to another to do. It is surprising how many architects are unaware of this provision and blithely omit work to give to others on a regular basis. What is even more surprising is the number of contractors who allow them to do it. There is no excuse for this behaviour. There are now two English cases to the same effect: *Vonlynn Holdings* v. *Patrick Flaherty Contracts Ltd* (1988); *AMEC Building Contracts Ltd* v. *Cadmus Investments Co Ltd* (1997).

The same principle applies with regard to the division of work within the contract. The original contract works will probably be divided up between work to be done and materials to be supplied by the main contractor on the one hand, and work to be done by nominated sub-contractors and materials to be supplied by nominated suppliers on the other. That division, once fixed by the contract, cannot be changed unilaterally by the employer acting through the architect. The main contractor has a right to do that work which is set out in the contract documents for him to do, and it cannot be taken away from him in order that it be done by a nominated sub-contractor. Conversely, if work is set out to be done by a nominated sub-contractor, the main contractor cannot be forced to do it instead; nor, for that matter, can he insist upon doing it himself: *North West Metropolitan Regional Hospital Board* v. *T.A. Bickerton & Son Ltd* (1970). Clause 13.1.3 of JCT 98 now spells out that restriction for the benefit of employers and architects who may not appreciate that it already exists, and therefore for the avoidance of the kinds of dispute over this point.

Clause 35.1.4 modifies the position, because it permits the nomination of a sub-contractor by agreement between the contractor and the architect on behalf of the employer. That clause would seem to empower the architect to agree with the contractor that work set out in the bills for the contractor to do will be done by a nominated sub-contractor. Without that express provision any such agreement would have had to be between the contractor and the employer himself, since it would amount to a 'variation of contract'. However, it is not clear what financial adjustment would be made to the contract sum. Should the value of the work as set out in the bills be deducted from the contract sum and the amount of the sub-contractor's account plus the normal level of the contractor's profit on PC sums be added, or should the contractor be entitled to the higher level of profit he would have anticipated had he done the work himself?

Valuations of variations to nominated sub-contract works (including approximate quantities in such works) are to be dealt with under the appropriate NSC/C provisions (clause 13.4.1.3).

If a contractor brings others onto the site to supplement a sub-contractor, already lawfully on site, without that sub-contractor's consent, the contractor is in breach of contract which may be repudiatory in nature so as to entitle the sub-contractor to leave site: *Sweatfield Ltd* v. *Hathaway Roofing Ltd* (1997).

There is an obligation in clause 13.5.7 upon the quantity surveyor to make a 'fair valuation' of 'any ... liabilities directly associated with a Variation', the valuation of which 'cannot reasonably be effected in the Valuation by the application of clauses 13.5.1 to 6'. This will include an obligation to make due allowance for factors such as the loss to the contractor involved where a variation to the work means that materials already properly ordered for the work as originally designed become redundant. It will also include the valuation of the effect of any instruction which does not require the addition, omission or substitution of work, i.e. obligations or restrictions.

Clause 13.5.5 requires the quantity surveyor, where the introduction of a variation results in other work not itself varied being executed under conditions other than those otherwise deemed to have been envisaged, to revalue that other work as if it had been varied. In practice this will mean that it must be revalued under clause 13.5.1.2 – that is, on the basis of the rates and prices in the contract bills against the appropriate items adjusted in respect of the changed conditions; but it may also be necessary for the quantity surveyor to make allowance for other factors such as consequential changes in preliminary items.

6.6 *Valuation of 'obligations and restrictions'*

Since the valuation of variations in the work to be executed under the contract or of work to be executed against provisional sums is comprehensively covered by clauses 13.5.1 to 13.5.5, clause 13.5.7, apart from the reference to 'liabilities directly associated with a Variation' **[6.5]**, must relate to the valuation of obligations or restrictions imposed by the employer or variations to obligations or restrictions already imposed in the contract bills as defined in clause 13.1.2. The quantity surveyor must make a fair valuation, but this is subject to the proviso that any effect of the variation on the regular progress of the works or of any direct loss and/or expense reimbursed to the contractor under any other provision of the contract must be excluded from the valuation. It is difficult to envisage what would remain to be valued in respect of such a variation other than its effect on the regular progress of the work. On a strict reading of its wording, therefore, the proviso appears to prevent the valuation of the effect of the removal of obligations or restrictions. The clause prevents allowance being made for any effect on the regular progress of the works – including any possible improvement in progress resulting from the removal of an obligation or restriction except to the extent that it discounts the contractor's entitlement under clause 26. So far as extension of time is concerned, the architect can take account of such removal under clause 25.3.2.

6.7   Rights of the contractor in respect of valuation

Subject to any agreement to the contrary between the employer and the contractor, the valuation of variations is the function of the quantity surveyor. The contractor has only one right, under clause 13.6: the opportunity of being present at the time of any measurement and of taking such notes and measurements as he may require. The contractor has no contractual right to be consulted, any more than has the employer or the architect. In theory, therefore, the quantity surveyor may simply notify the contractor of his intention to take measurements, but if the contractor has a prior appointment the quantity surveyor, having given him the opportunity of being present, may proceed without him. When valuing the results of such measurement the quantity surveyor has no obligation to consult the contractor at all, but may proceed without him and at the end of the contract, as required by clause 30.6.1.2, may simply present the contractor with the statement of all the adjustments to the contract sum which would include a summary of the variation valuations, possibly without even measurements attached. The contractor would then either have to accept it or, in due course, to refer the matter to arbitration when it became enshrined in an architect's certificate.

In practice, quantity surveyors almost invariably measure and value variations in full consultation and, as far as possible, in agreement with the contractor, so that the adjustment of the contract sum will be an agreed document at least insofar as the variations are concerned and a possible area of dispute will be removed. It is in everyone's interests that this is done. However, the quantity surveyor has the authority under the contract to proceed unilaterally if necessary.

6.8   *Payment in respect of variations*

A valuation under clause 13.4.1.1 must be given effect by adding to or deducting from the contract sum (clause 13.7). This form of words means that the amount of the valuation must be taken into account in the computation of the next interim certificate (clause 3). Read strictly, this could pose a difficulty if the amount was taken into account in the next interim certificate after the valuation had been made, but before the work had been carried out. The quantity surveyor could ensure that such valuations are not completed until the work is executed, but that is avoiding the issue. However, the position is correctly stated in clause 30.2.1.1, which states that there shall be included in interim certificates 'the total value of the work properly executed by the Contractor'. If, unusually, the valuation is made before the work is properly executed, it may be 'taken into account' in the sense of being considered, but it would not be included because not properly executed. If the formal valuation has not been made by the time the work has been properly executed, an allowance should be made for it in the next interim certificate.

6.9   *The allowance of preliminaries*

Clause 13.5.3 sets out particular requirements which must be satisfied if the work is valued under clause 13.5.1 and 13.5.2. The overall criterion is that measurement is to be in accordance with the same principles which applied to the preparation of the bills of quantities (clause 13.5.3.1). The quantity surveyor must take into account factors other than the prices set out in the contract bills against individual items or his fair valuation of measured items when valuing omissions, additional and substituted work. Clause 13.5.3.2 requires him to make allowance for 'any percentage or lump sum adjustments in the Contract Bills', that is, any such percentages or lump sums must be applied pro rata to all prices for measured work.

The quantity surveyor must also make 'allowance, where

appropriate . . . for any addition to or reduction of preliminary items of the type referred to in the Standard Method of Measurement,' (clause 13.5.3.3). It should be noted that the clause does not actually bind the quantity surveyor to use the rates and prices set out in the bills against such items, but simply to make allowance for any 'addition to or reduction of' such items.

Attention should be drawn to the limitation on the duty and power of the quantity surveyor with regard to the valuation of variations which is set out in the proviso at the end of clause 13.5, that is that 'no allowance shall be made under clause 13.5 for any effect upon the regular progress of the Works or for any direct loss and/or expense for which the Contractor would be reimbursed by payment under any other provision in the Conditions' – the reference being, of course, to clause 26. In making any allowances in respect of preliminary items, therefore, the quantity surveyor must stop short of making allowance for the effect of the variation in question on regular progress of the works.

6.10   Clause 13A quotations

Clause 13A is needlessly complicated by the fact that it involves employer, architect and quantity surveyor in the procedure.

If the contractor has received sufficient information with any architect's instruction, he must provide a '13A Quotation', including clause '3.3A Quotations' in respect of variations to nominated sub-contract work where appropriate, not later than 21 days from the date of receipt of the instruction. Although the instruction is to be issued by the architect, the quotation must be sent to the quantity surveyor where it is open for acceptance by the employer for seven days. Unusually, it appears that the contractor cannot withdraw the quotation before acceptance as he could in the course of ordinary negotiations, because in this instance he is bound by the contract terms to keep his offer open. The quotation must contain:

- The value of the adjustment to the contract sum
- Adjustment to the contract period including fixing a new, possibly earlier, completion date
- The amount of loss and/or expense
- The cost of preparation of the quotation.

If the architect specifically states, the contractor must also include:

- Additional resources required
- A method statement.

The employer's role is important, probably because the contractor will be expected to quote where it is likely that the instruction will have some significant effect on the contract in terms of additional expenditure or time. It is for the employer to accept the quotation or otherwise and to notify the contractor in writing and, if he accepts, the architect must confirm the acceptance in writing to the contractor (clause 13A.3). The purpose of this acceptance is that the architect can formally confirm that the contractor is to proceed, that the adjustment to the contract sum can be made, that a new date for completion (if applicable) can be fixed including revised sub-contract periods for any nominated sub-contractors and that the contractor is to accept any clause '3.3A Quotation'. The provision, that if the employer does not accept, the architect must either instruct that the variation is to be carried out and valued under clause 13.4.1 or instruct that the variation is not to be carried out, is remarkable in one particular. There seems to be no provision for the employer or the quantity surveyor on his behalf to negotiate on the quoted price. It is either to be accepted or rejected.

6.11   *The contractor can offer a quotation – alternative A*

Under clause 13.4.1.1, all variations are to be valued under alternative A if they fall into the following categories:

- Variations instructed or sanctioned (under clause 13.2.4) by the architect
- All work which the contract states is to be treated as a variation
- All work carried out by the contractor under instructions from the architect dealing with the expenditure of a provisional sum included in the contract
- All work carried out where approximate quantities have been included in the bills of quantities.

However, if any of such work relates to nominated sub-contract work, it is to be valued under the provisions of NSC/C.

6.12   *The quantity surveyor's dominant role*

It is solely the responsibility of the quantity surveyor to determine the price to be paid or allowed in respect of a variation unless the employer and the contractor agree otherwise. The architect has no direct authority in the matter. It follows that if the architect included, in an instruction requiring a variation, any purported direction as to how it should be valued, such as 'the work executed against this instruction is to be valued as daywork', this would be of no effect and the quantity surveyor not only should, but must, ignore it and use his own judgement about the way in which the work should be valued under the terms of the contract. Whether such an instruction issued by the architect is rendered void is not clear; it may remain valid except for the part regarding valuation, which is an instruction which the architect is not empowered to give. Of course, the architect has the power (and the duty) to certify what he believes to be due in accordance with the contract and it is probable that he has the power to make an adjustment if he particularly disagrees with the quantity surveyor. The quantity surveyor's function is to value the work as set out in the contract. The contract provides that the quantity surveyor is to value the work in accordance with the provisions of clauses 13.5.1 to 13.5.7.

6.13   A fair and reasonable rate

With respect to what may be considered 'fair rates and prices' for valuation under clause 13.5.1.3, it is suggested that the word 'fair' must be read in the context of the contract as a whole. A 'fair' price for varied work in a contract where the prices in the bills are 'keen' will be a similarly keen price: *Phoenix Components* v. *Stanley Krett* (1989). The quantity surveyor should determine his 'fair rates and prices' on the basis of a reasonable analysis of the contractor's pricing of the items set out in the bills, including his allowances for head-office overheads and profit.

6.14   Daywork

If additional or substituted work cannot 'properly', i.e. by reference to SMM7, be valued by measurement, it is to be valued as 'daywork', that is on the basis of prime cost plus percentages as set out in clause 13.5.4. Clearly, while this may be a satisfactory method of valuation to the contractor since it ensures that he will, at least, recover the cost to him of the work (subject to the limitations imposed by the relevant 'Definition of Prime Cost' defined in the clause) plus percentages to cover supervision, overheads and profit, it is not necessarily a satisfactory method for the employer since it imposes no incentive on the contractor to work efficiently. It should, therefore, be used sparingly by the quantity surveyor and only if measurement is quite impossible.

The machinery for submission of daywork vouchers, particularly the timing, is also unsatisfactory. The requirement that the vouchers should be 'delivered for verification to the architect or his author-ised representative not later than the end of the week following that in which the work has been executed' is quite unworkable. If the architect or his representative is to verify what is set out on the voucher, i.e. to vouch for the truth of it, it is wholly unreasonable to expect him to be able to do so when a voucher for work executed on the Monday of one week does not have to be delivered to him for that purpose until Friday of the following week – an interval of 11 days. In fact, it will be difficult enough if submitted on the day following the carrying out of the work. Only if the architect or his representative is permanently on site will he be able to verify, because verification is simply checking that time and materials were spent, not whether they should have been spent.

It is suggested that the most sensible way to deal with the prob-lem of 'verification' of daywork vouchers is for the contractor to give advance notice to the architect of his intention to keep daywork records of a particular item of work; for the architect himself to attend the site or, if he is unable to do so, to nominate the clerk of works to act as his 'authorised representative' for that purpose and to take his own records of the time spent and materials used; and for the vouchers to be submitted for verification at the end of each day. In that way, at least, the quantity surveyor can be reasonably certain that the vouchers represent an accurate record of time and materials. If this system is to work properly, it requires the quantity surveyor to notify the contractor in advance of his intention to value using daywork. Of course, he is still under no obligation to accept daywork as the method of valuation if, in his opinion, the work can properly be measured.

This is the only reference to the possible existence of the archi-tect's representative in the entire contract. It is generally assumed that the clerk of works will take that role. Reference to clause 12, however, will show that the clerk of works, far from being a representative of the architect, is an 'inspector on behalf of the Employer', and that he therefore has no authority to verify daywork vouchers unless the architect specifically gives him that authority.

Signing the sheets normally indicates verification. Often 'For record purposes only' is added. However, where dayworks is to be the method of valuation in any particular case, the addition of those words has little practical value and certainly does not prevent the contents of the sheets being used for calculation of payment: *Inserco v. Honeywell* (1996). In these circumstances it appears that the quantity surveyor has no right to substitute his own opinion for the hours and other resources on the sheets: *Clusky (t/a Damian Construction) v. Chamberlain* (1995).

CHAPTER SEVEN

LIQUIDATED DAMAGES AND EXTENSION OF TIME

7.1 Provisions regarding liquidated damages

Liquidated and ascertained damages are examined in some detail in *Powell-Smith & Sims' Building Contract Claims* (3rd edition) (1998). What follows is necessarily a brief summary of the position. In order to recover damages in matters involving breaches of contract it is necessary to prove that the defendant had a contractual obligation to the plaintiff, that there was a failure to fulfil the obligation wholly or partly and that the plaintiff suffered loss or damage thereby. Very often it is clear that there is damage, but it is difficult and expensive to prove it. To avoid that situation, the parties may decide when they enter into a contract, that in the event of a breach of a particular kind the party in default will pay a stipulated sum to the other. This sum is termed liquidated damages.

The terms 'liquidated damages' and 'penalty' are commonly used as though they were interchangeable. They are totally different. Liquidated damages are a genuine attempt to predict the damages likely to flow as a result of a particular breach, but a penalty is a sum which is not related to probable damages, but rather stipulated *in terrorem*: *Cellulose Acetate v Widnes Foundry* (1933). The courts will enforce the former, but not the latter even though the sum may have been agreed by both parties and inserted in the contract.

Provided the contractor is able to enter on the site on the date stipulated for possession and thus to commence building work, he must finish by the completion date. If he fails to complete, the employer may recover such damages as he can prove were a direct result of the breach.

In certain instances, it may be difficult to decide which damages directly and naturally flow from the breach and which damages do not so flow but depend on special knowledge which the contractor had at the time the contract was made. For more than a hundred years it has been the practice in the building industry to include a provision for liquidated damages in building contracts to avoid these difficulties. The provision is generally expressed that the contractor must pay a certain sum to the employer for every week by which the original completion date is delayed. That sum must represent a genuine pre-estimate of the loss which the employer is likely to suffer.

7.2   *When are liquidated damages really a penalty?*

Whether a sum is to be considered liquidated damages or a penalty can sometimes be a tricky question. Guidance was set out by Lord Dunedin in *Dunlop Pneumatic Tyre Co Ltd* v. *New Garage & Motor Co Ltd* (1915). Since the case of *Kemble* v. *Farren* (1829), the courts have paid little attention to the terminology adopted by the parties, sometimes holding that 'liquidated damages' were penalties, and in other cases, holding that sums stated as 'penalties' were in fact liquidated damages: *Ranger* v. *Great Western Rail Co* (1854).

A sum may be liquidated damages although it is not a genuine pre-estimate; for example if the sum is agreed at a lower figure. The decision whether a sum is liquidated damages or penalty will hinge not only on the terms of a particular contract, but also on the inherent circumstances of that contract. The terms and inherent circumstances to be considered are those existing at the time the contract was made, not when the term was breached. This is important when considering whether a sum is a genuine pre-estimate of loss. In looking at a sum, it should be considered in the worst possible light just as, if there are several possible breaches, 'the strength of the claim must be taken at its weakest link': *Dunlop Pneumatic Tyre Co Ltd* v. *New Garage & Motor Co Ltd* (1915). Therefore, if a sum would not normally be considered a penalty, but under certain circumstances it would be penal, then it is to be treated as penal in its entirety and the court will not sever any part.

Lord Dunedin proceeded to set out tests which could prove helpful or even conclusive:

'(a) It will be held to be a penalty if the sum stipulated for is extravagant and unconscionable in amount in comparison with the greatest loss which could conceivably be proved to have followed from the breach.
(b) It will be held to be a penalty if the breach consists only in not paying a sum of money, and the sum stipulated is a sum greater than the sum which ought to have been paid.

(c) There is a presumption (but no more) that it is a penalty when "a single lump sum is made payable by way of compensation, on the occurrence of one or more or all of several events, some of which may occasion serious and others but trifling damages".
(d) It is no obstacle to the sum stipulated being a genuine pre-estimate of damage that the consequences of the breach are such as to make precise pre-estimation almost an impossibility. On the contrary, that is just the situation when it is probable that pre-estimated damage was the true bargain between the parties.'

From this it is clear that the two important points are the extent to which an accurate pre-estimate of loss can be carried out, and the existence of different events each of which are said to give rise to liquidated damages. The decision of the Privy Council of the House of Lords in *Philips Hong Kong Ltd* v. *Attorney General of Hong Kong* (1993), is significant. The Law Lords held that hypothetical situations cannot be used to defeat a liquidated damages clause. The court will take a pragmatic approach. A sum will be classed as liquidated damages if it can be said of it that it is a genuine pre-estimate of the loss or damage which would probably arise as a result of the particular breach. The figure inserted in the contract must be a careful and honest attempt to accurately calculate the loss or damage which will be suffered and it must be a pre-estimate in the sense that it must be an estimate at the time the contract is made, not at the time of the breach: *Public Works Commissioner* v. *Hills* (1906).

The courts seem to be willing to accept sums which are greater than that which would constitute a genuine pre-estimate in certain limited circumstances: *The Angelic Star* (1988). The court appears to have looked on a repayment provision as a form of liquidated damages.

7.3   Are actual damages an option?

Can a party to a contract containing a liquidated damages clause sue for actual damages suffered or is he is restricted to the sum expressed as liquidated damages? Where parties enter into a contract, it must be assumed that they know what they are doing and that the contract is an expression of their intentions: *Liverpool City Council* v. *Irwin* (1976). If parties agree that in the event of a particular kind of breach liquidated damages are payable by the party in breach, that agreement will be upheld by the courts and they will be

allowed no other or alternative damages but the damages liquidated in the contract. The sum expressed as liquidated damages
was held to be exhaustive of the remedies available to the plaintiff
for late completion in *Temloc Ltd* v. *Errill Properties Ltd* (1987) where
the amount of liquidated damages was stated to be 'Énil'. It was
held that the parties had agreed that, in the event of late completion,
no damages should be applied. Even if a rate had been stated, the
court considered that the rate would have represented an exhaustive agreement as to damages which were or were not to be payable
by the contractor in event of his failure to complete on time. *M.J.
Gleeson plc* v. *Taylor Woodrow Construction Ltd* (1989) is a case to
similar effect.

   This principle should be distinguished from the situation where
the defendant is in breach of two or more obligations, for one of
which the stipulated remedy is liquidated damages and for the
other(s) the remedy is to sue for unliquidated damages. A related
situation is where there is but one breach which gives rise to a loss
which may be said to trigger a remedy in liquidated damages and a
separate kind of loss for which other damages are appropriate. In *E.
Turner & Sons Ltd* v. *Mathind Ltd* (1986), a number of flats were to be
completed in stages and there was a final completion date for the
whole development. Liquidated damages were stipulated only for
failure to meet the final completion date. Although expressed
obiter, it was the court's view that the liquidated damages clause,
standing alone, was not an effective exclusion of any right to
damages for earlier breaches of the obligations to meet intermediate
dates.

7.4   Can liquidated damages be recovered if it can be shown that there is no loss?

Whether a party is entitled to recover the amount specified as
liquidated damages if the damage actually suffered is less than the
amount or nothing at all, is a question which arises with surprising
frequency. Indeed, is he able to recover liquidated damages though
it can be demonstrated that he has actually gained from the breach?
It is settled that a party can recover liquidated damages without
being put to proof of actual loss: *Clydebank* v. *Castenada* (1905). In
some instances the actual loss will be greater and in others less than
the sum in the contract. Indeed, it follows that in certain instances
there will be no loss whatever. The sum named as liquidated
damages in the contract is recoverable whether or not the employer

can prove that he has in fact suffered any loss or damage as a result of the breach: *Crux* v. *Aldred* (1866).

The principle was applied in *BFI Group of Companies Ltd* v. *DCB Integration Systems Ltd* (1987). On an appeal from the award of an arbitrator, it was found that, although the plaintiffs had suffered no actual loss as a result of being unable to use two vehicle bays, because they had, in any event, to execute fit-out works after possession before being able to attract revenue, they were entitled to liquidated damages. The form of contract was on MW 80 terms and provided for the payment of liquidated damages if completion was delayed beyond the completion date. The plaintiffs were given possession on the extended date for completion although the arbitrator found that practical completion had not taken place. Had they not been given possession, they would have been obliged to wait until practical completion was certified before being able to execute the fit-out works. Unlike other forms of contract, such as JCT 98 or IFC 98, in MW 98 (and in MW 80 before it) there is no provision for possession of part of the Works before practical completion and the possession granted to the plaintiff in this case was a concession. Therefore, the plaintiffs were able to carry out work during the period within which they were receiving liquidated damages. In such circumstances, this may have a considerable advantage to the plaintiff.

To produce a genuine pre-estimation of loss is not easy. The employer may have little idea how much loss he may suffer if the building is not completed by the due date, particularly if the contract period is to be counted in years rather than months. As long ago as 1787, it was held that the fact that damages when the contract was made were difficult to quantify or estimate, as in the case of a church, does not prevent a sum named in the contract being liquidated damages: *Fletcher* v. *Dycke* (1787).

Although it has been held that liquidated damages are especially suited to situations where precise estimation is almost impossible (*Dunlop Pneumatic Tyre Co Ltd* v. *New Garage & Motor Co Ltd* (1915)), the employer should do his best to calculate as accurate a figure as possible. The employer should include every item of additional cost which he predicts will flow directly from the contractor's failure to complete on the due date; that is, the damages recoverable under the first limb of the rule in *Hadley* v. *Baxendale* (1854). It seems that the sum can be increased to include amounts which would normally only be recoverable under the second limb if the employer can show that special circumstances were involved: *Philips Hong Kong Ltd* v. *Attorney General of Hong Kong* (1993). It remains unclear whether, in the case of liquidated damages, the special circumstances must be known to the contractor when the contract is made. It seems appropriate to reveal such circumstances at tender stage.

An employer will very often reduce such a figure in order to make the proposed damages more palatable to prospective tenderers. Some local authorities and other public bodies make use of a formula calculation which basically depends upon a percentage of the capital sum. Whether that would constitute liquidated damages will depend on the precise circumstances and particularly the difficulty with which a precise calculation could be made.

A practical problem concerns the employer's position if liquidated damages are held to be a penalty. Is he restricted to recovery of such amount as he can prove, up to but not greater than the amount of the sum held to be penal? In an early judgment in the Court of Appeal, Lord Justice Kay traced the effect of courts of equity on sums stipulated as penalties and noted that if the actual damages could easily be estimated, 'the penalty would be cut down and the actual damage suffered would be assessed': *Law* v. *Redditch Local Board* (1892). The Supreme Court of Canada has said:

> 'If the actual loss turns out to exceed the penalty, the normal rules of enforcement of contract should apply to allow recovery of only the agreed sum. The party imposing the penalty should not be able to obtain the benefit of whatever intimidating force the penalty clause may have in inducing performance, and then ignore the clause when it turns out to be to his advantage to do so. A penalty clause should function as a limitation on the damages recoverable, while still be ineffective to increase damages above the actual loss sustained when such loss is less than the stipulated amount.': *Lorna P. Elsley* v. *J.G. Collins Insurance Agencies Ltd* (1978).

This probably represents the modern approach to this problem, but it is not clear whether it necessarily represents the position following the failure of the liquidated damages clause for every reason. A penalty is always a sum which is extravagant in relation to the damages likely to be incurred, but liquidated damages can operate as a limitation on damages: *Cellulose Acetate Silk Co Ltd* v. *Widnes Foundry (1925) Ltd* (1933). In the case of liquidated damages in a building contract, no default on the part of the contractor can prevent the application of the clause. Only by a default on the part of an employer can the clause fail. A contractor who enters into a contract with an employer, which includes a relatively small sum for liquidated damages, will have a valuable advantage. The employer will be equally and oppositely disadvantaged, but both parties will have agreed on the arrangement as part of the distribution of risk inherent in that particular contract.

If the employer is so minded, it is possible for him to disable the liquidated damages clause by failing to grant an extension of time in appropriate circumstances and then he would be entitled to claim whatever amount of unliquidated damages he could prove: *Peak Construction (Liverpool) Ltd* v. *McKinney Foundations Ltd* (1970). If the sum stipulated in the contract is not a ceiling on what can be claimed in those circumstances, it would be open to the employer to effectively alter the distribution of risk and, as a result of his own default, be entitled to a greater sum in damages than if he had properly performed his part of the bargain. Purely on the principle that a party cannot profit by his own contractual breach to the detriment of the other party, there is a strong argument that the liquidated damages sum must be a ceiling on recovery: *Alghussein Establishment* v. *Eton College* (1988).

7.5 Procedure for recovery of liquidated damages

Three conditions must be satisfied before the employer is entitled to liquidated damages. First, there must be a failure by the contractor to complete the works by the date for completion specified or within any extended time. Secondly, the architect must have performed his duties in deciding on extensions of time under clause 25: *Token Construction Co Ltd* v. *Charlton Estates Ltd* (1973). The third condition to be observed is that the architect must have issued a certificate under clause 24.1 to the effect that the contractor has failed to complete by the completion date, usually referred to as the 'non-completion certificate'. He can issue the certificate at any time prior to the issue of the final certificate. In practice, of course, most architects would be well-advised to issue the certificate immediately the completion date has passed in order to allow the employer the maximum possible time and maximum available funds for deduction of liquidated damages. An employer may be able to bring an action against an architect who delays the issue of the certificate until just before the issue of the final certificate if by that time it is impossible to recover the liquidated damages. Once the architect has issued the final certificate under clause 30.8, if no notice of adjudication or arbitration has been given by either party in accordance with clause 30.9, the architect becomes *functus officio* and is thereby excluded thereafter from issuing any valid certificate under clause 24: *H. Fairweather Ltd* v. *Asden Securities Ltd (1979)*.

The architect is obliged to issue the certificate of non-completion if the contractor has failed to complete by the due date. If the architect fixes a new date for completion after the issue of the certificate, the fixing of a new date is said to cancel the existing certificate and the architect must issue a further certificate. The clause refers to 'such further certificate ... as may be necessary', because if the architect fixes a new date which is the same as, or later than, the date the contractor actually completes the works, a further certificate is unnecessary.

Clause 24.2.1 introduces two further conditions precedent if the employer wishes to deduct them from future payments or if he wishes the contractor to pay him. The employer must inform the contractor in writing that he may require him to pay or that the employer may deduct liquidated damages, then the employer must give a written notice to the contractor that he requires the payment (clause 24.1.1) or that he intends to deduct the amount of liquidated damages (clause 24.1.2). This second of the employer's required notices must be given no later than five days before the final date for payment in a certificate. The contract text misleadingly refers to the necessity of serving the notice no later than five days before the final date for payment of the *final certificate*. However, that would only be appropriate if it was proposed to deduct the amount from payment of the final certificate when it has already been said that there may be little from which to deduct.

Clause 24.2.3 provides that the employer need only serve one notice under clause 24.2.1 requiring payment. It remains effective, despite the issue of further non-completion certificates, unless the employer withdraws it. Since the decision to require payment or to deduct liquidated damages rests with the employer, it is unlikely that he would ever, in practice, withdraw the notice. If he decided not to deduct damages, he would simply let the matter rest. The precise form to be taken by the employer's written requirement under clause 24.2.1 was finally clarified by a decision of the Court of Appeal in *J.J. Finnegan Ltd* v. *Community Housing Association Ltd (1995)* where the court held that the employer's written requirement was a condition precedent to the deduction of liquidated damages. Only two things must be specified in the requirement and they are:

- Whether the employer is claiming a payment or a deduction of the liquidated damages; *and*

- Whether the requirement relates to the whole or part of the total liquidated damages.

The amount which the employer may deduct is to be calculated by reference to the rate stated in the appendix. The employer is free to reduce the rate, but not to increase it. Clause 24.2.1 makes clear that the employer need not wait until practical completion before deducting liquidated damages. He may start to deduct them as soon as the clause 24.1 certificate has been issued, the requirement for payment made and the withholding notice served. In practice, such deductions usually commence from the first payment after the contractor falls into culpable delay.

Clause 24.2.2 provides that if there is a later completion date fixed under the contract provisions, the employer is obliged to repay any money deducted as liquidated damages up to the later completion date. It should be noted that no interest is payable by the employer, despite the oft misquoted decision in *Department of the Environment for Northern Ireland* v. *Farrans (Construction) Ltd* (1982).

7.6 Time at large and loss of right to liquidated damages

It is essential that a date for completion must be inserted in the contract if the parties intend that liquidated damages are to be payable if the contractor fails to complete the works. That is because there must be a definite date from which to calculate liquidated damages: *Miller* v. *London County Council* (1934). There is an implied term in every contract that the employer will do all that is reasonably necessary to co-operate with the contractor (*Luxor (Eastbourne) Ltd* v. *Cooper* (1941)) and, perhaps more importantly, that he will not prevent him from performing: *Cory Ltd* v. *City of London Corporation* (1951).

An implied term that neither party will do anything to hinder or delay performance by the other was upheld as generally applicable to building contracts in *London Borough of Merton* v. *Stanley Hugh Leach Ltd* (1985). If the employer does hinder the contractor, he cannot insist that the contractor finishes his work by the contractual date for completion. In *Dodd* v. *Churton* (1897), it was held that an employer cannot recover liquidated damages if he prevents the contractor completing within the stipulated time. Very clear words will be needed in order to bind a contractor to a completion date if the employer is the cause of the delay. This principle is now well established: *Percy Bilton* v. *Greater London Council* (1982). If there is no agreed contractual mechanism for fixing a new date for completion, no such new date can be fixed and the contractor's duty then will be to complete the works within a reasonable time: *Wells* v. *Army & Navy Co-operative Society Ltd* (1902). Time is then said to be 'at large'.

All modern standard form building contracts have a clause enabling the employer or his agent to fix a new completion date after the employer has caused delay to the contractor's progress. All standard forms have clauses permitting the extension of time although not all of them are entirely satisfactory. Even where a building contract contains terms providing for extension of the contract period, time may become at large, either because the terms do not properly provide for the delaying event or because the architect has not correctly operated the terms. Regrettably, the latter is more common.

JCT 98 largely, although not entirely, fulfils the requirement that extension of time clauses should be drafted so as to include for all delays which may be the responsibility of the employer. Then, if the employer, either personally or through the agency of his architect, hinders the contractor in a way which would otherwise render the date for completion ineffective, the architect will have the power to fix a new date for completion and thus preserve the employer's right to deduct liquidated damages. Some other forms of contract give blanket coverage by allowing the architect to give an extension of time for any action, omission or default of the employer which causes delay.

The JCT series of contracts (other than MW 98) favour a list of events giving grounds for extension of time. Because the architect's power to give an extension of time is circumscribed by the listed events, there is a danger that the employer may delay the works in a way which does not fall under one of the events. In such a case, time would be at large. For example, the 1963 edition of the JCT Standard Form did not include power for the architect to extend time for the employer's failure to give the contractor possession of the site on the due date. An employer's failure in this respect resulted in time becoming at large and the contractor's obligations being to complete the works within a reasonable time. This although it was acknowledged by the court that the contractor had himself subsequently contributed to the delay: *Rapid Building Group Ltd* v. *Ealing Family Housing Association Ltd* (1984). Surprisingly, it has been held that the architect has the power to give an extension of time if the employer causes further delay when the contractor is already in delay through his own fault after the contract date for completion: *Balfour Beatty Ltd* v. *Chestermount Properties Ltd* (1993).

If the extension of time clauses are properly drafted, but the architect operates them incorrectly, time may become at large. For example, if the architect was late in providing necessary information to the contractor, but failed to give any extension of time. This is a clear case of the architect not taking advantage of the available mechanism. Another example is where the contract provision sets out a timetable (such as the 12 week review period under clause 25.3.3) within which the architect must operate to give an extension of time. If he fails to observe the timetable, his power to give an extension will end and time will become at large. It has been said that such time periods are not mandatory, but simply directory on the authority of the Court of Appeal in *Temloc Ltd* v. *Errill Properties Ltd*. It appears the court in *Temloc*, in making that observation, were interpreting the provisions *contra proferentem* the employer who sought to rely on them. The employer had stipulated '£nil' as the figure for liquidated damages and the Court of Appeal held that this meant that the parties had agreed that if the contractor finished late, no liquidated damages would be recoverable by the employer. The court held that the employer could not claim unliquidated damages. The contract (which was the 1980 edition of the JCT standard form with an identical review clause) provided that after practical completion the architect must, within 12 weeks, confirm the existing date for completion or fix a new date. The architect exceeded the 12 weeks and the employer contended that the liquidated damages clause could be triggered only if the architect carried out his duty at the right time. Therefore, the employer could claim unliquidated damages. It was apparently in this context that the court, in a view which is probably *obiter* in any event, suggested that the time period was not mandatory. The court recognised that the architect is the employer's agent. Had the employer's argument succeeded, it would have been contrary to the established principle that a party to a contract cannot take advantage of his own breach. It is to be regretted that the court did not make its reasoning clear.

The employer may be estopped (prevented) from exercising his right to recover liquidated damages if he tells the contractor before the date for completion is exceeded, that he will not be seeking damages and if the contractor then acts in reliance on the statement: *London Borough of Lewisham* v. *Shephard Hill Engineering* (2001). The question is whether it would be inequitable or unjust to subsequently allow the right to be enforced.

7.7   *Employer's failure to nominate*

In *Percy Bilton* v. *Greater London Council* (1982), the plaintiffs were many weeks behind with their own work and the nominated sub-contractor for mechanical services was not able, for this reason, to start work for 22 weeks. Thirty-six weeks later the nominated sub-contractor withdrew from the site and went into liquidation on 31 July 1978. The same day another contractor for mechanical services was engaged on a daywork basis. By 14 September 1978, the mechanical services contractor employed on a daywork basis became the nominated sub-contractor. The architect extended time for completion under JCT 63 clause 23(f) for this delay in renomination.

On 4 February 1980, the architect issued a certificate under JCT 63 clause 22 that the contract ought reasonably to have been completed by 1 February 1980. Thereafter the GLC began to deduct liquidated damages, amounting to £97,543 in all, from the sums certified in interim certificates.

Then commenced a long legal battle. Eventually the case went to the House of Lords on the question of responsibility for the delay and the duty to renominate.

The House of Lords noted that the sub-contractor's employment had come to an end by notice given by a receiver that labour would be withdrawn and this, it was held, constituted a repudiation of the sub-contract which Percy Bilton had accepted. It was common ground that the delay caused to the contractor fell into two parts:

- Arising from the withdrawal of the sub-contractor
- From the failure of the GLC to nominate a replacement sub-contractor with reasonable promptness.

For the latter period, for which the GLC accepted responsibility, Percy Bilton had been given an extension of 14 weeks. A passage in the judgment of Lord Justice Stephenson in the Court of Appeal was expressly approved:

'Insofar as delay was caused by the departure of [the sub-contractor] ... it was a delay which was not within the provision of clause 23.

Therefore the plaintiff was not entitled to any extension of time in respect of it, with the result not that time became at large but that ... the date for completion remained unaffected.'

In the House of Lords, Lord Fraser memorably said:

'(1) The general rule is that the main contractor is bound to complete the work by the date for completion stated in the contract. If he fails to do so, he will be liable for liquidated damages to the employer.

(2) That is subject to the exception that the employer is not entitled to liquidated damages if by his acts or omissions he has prevented the main contractor from completing his work by the completion date – see for example *Holme* v. *Guppy* (1838) 2 M&W 387, and *Wells* v. *Army and Navy Co-operative Society* (1902) 86 LT 764.

(3) These general rules may be amended by the express terms of the contract.

(4) In this case the express terms of clause 23 of the contract do affect the general rule. For example, where completion is delayed (a) by *force majeure*, or (b) by reason of any exceptionally inclement weather, the architect is bound to make a fair and reasonable extension of time for completion of the work. Without that express provision, the main contractor would be left to take the risk of delay caused by *force majeure* or exceptionally inclement weather under the general rule.

(5) Withdrawal of a nominated sub-contractor is not caused by the fault of the employer, nor is it covered by any of the express provisions of clause 23. Paragraph (g) of clause 23 expressly applies to "delay" on the part of a nominated sub-contractor but such "delay" does not include complete withdrawal . . .

(6) Accordingly, withdrawal falls under the general rule, and the main contractor takes the risk of any delay directly caused thereby.

(7) Delay by the employer in making the timeous nomination of a new sub-contractor is within the express terms of clause 23(f) and the main contractor, the appellant, was entitled to an extension of time to cover that delay. Such an extension has been given.'

The exact terms of JCT 63 clause 23(f) were:

'by reason of the Contractor not having received in due time necessary instructions, drawings, details or levels from the Architect for which he specifically applied in writing on a date which having regard to the Date of Completion stated in the appendix to these Conditions or to any extension of time then fixed under this clause or clause 33(1)(c) of these Conditions was neither unreasonably distant from nor unreasonably close to the date on which it was necessary for him to receive the same . . .'

Nowhere in JCT 63 did it suggest that the main contractor is under an obligation to apply to the architect *in writing* for a renomination. In fact it was clearly held by the House of Lords in *North West Metropolitan Regional Hospital Board* v. *T.A. Bickerton* (1970) that the obligation was on the employer to provide a replacement sub-contractor. But in this case, Percy Bilton did apply *in writing* for the architect's instructions. If they had not done so, would the architect have been powerless under clause 23(g) to extend time for the employer's unreasonable delay in renomination? If he had no such power under the contract, time must have been at large, since it was common ground that in regard to the second period of delay, the employer was at fault. Did that one letter cost the contractor what is believed to be £2 million in liquidated damages **[7.11.1]**?

The basic scheme of nomination, involving NSC/W, seeks to give the employer who has lost his right to liquidated damages against the contractor recourse in damages against the nominated sub-contractor for breach of warranty contained in NSC/W clause 3.3.2. But that warranty must be read in the light of the provision of the contractor/sub-contractor contracts which in NSC/C clause 2.3 requires the architect to consent to the contractor giving an extension of time to the nominated sub-contractor for a whole string of relevant events set out in NSC/C, clause 2.6 plus the contractor's own default.

Two things are clear: if the employer wishes to recover lost liquidated damages he will have to bring an action against the nominated sub-contractor and he will have extreme difficulties in proving a breach of the warranty.

The same attempt to exact a warranty from the nominated suppliers is to be found in the standard form of tender. Clause 25.4.7 of JCT 98 qualifies delay on the part of the sub-contractor by using the words: 'which the contractor has taken all practical steps to avoid or reduce'. It is the sub-contractor's delay to which this subordinate clause applies.

7.8   Employer's failure to give possession of site

There is an implied term in every building contract that the employer will give possession of the site to the contractor within a reasonable time, i.e. in time to enable the contractor to complete the works by the contractual completion date. Under JCT 98, there is specific provision for the contractor to be given possession on the date specified in the Appendix.

Any failure by the employer to give possession on the date named is a breach of contract. Since default in giving possession is a breach of a major term of the contract, prolonged failure to give possession, and acceptance by the contractor of the employer's breach, entitle the contractor to repudiate the contract and to sue for damages which would include the loss of the profit that he would have earned if the contract had been completed: *Wraight Ltd* v. *P.H. & T. Holdings Ltd* (1968) . Contractors seldom wish to take such a drastic course and, therefore, they may decide to treat the breach as a minor matter only and to claim damages at common law for any loss actually incurred. At the very least, the contractor is entitled to damages for breach: *London Borough of Hounslow* v. *Twickenham Garden Developments Ltd* (1970).

It is sometimes argued that *Twickenham Garden* is authority for what is sometimes referred to as 'sufficient possession' and that, therefore, the employer need give only that degree of possession which is necessary to enable the contractor to carry out work. In *Freeman & Son* v. *Hensler* it was stated:

> 'I think there was an implied condition on the part of the defendant that he would hand over the land to the plaintiffs to enable them to carry out what they had contracted to do, and that it applied to the whole area.'

This concerned a contract in which nothing was said about possession. The court considered the matter so important that they were prepared to imply a term that possession of the whole site must be given. The commentary to *The Queen* v. *Walter Cabot Construction* (1975) in *Building Law Reports* contains the following:

> 'English standard forms of contract, such as the JCT Form, proceed apparently on the basis that the obligation to give possession of the site is fundamental in the sense that the contractor is to have exclusive possession of the site. It appears that this is the reason why specific provision is made in the JCT Form for the employer to be entitled to bring others on the site to work concurrently with the contractor for otherwise to do so would be a breach of the contract. No such right could be implied, at least on the wording of the standard form. This right is circumscribed since if completion of the works is delayed by the activities of those engaged by the employer or if the progress of the work is

materially affected then the contractor may be entitled to an extension of time or compensation or both, as the case may be ...'

This appears to be a correct view.

Whether or not the contractor has been given sufficient possession is a matter of fact. In *The Rapid Building Group Ltd* v. *Ealing Family Housing Association Ltd* (1984), which arose under a contract in JCT 63 form, at the time when, by clause 21, the defendants were bound to give the plaintiffs possession of the site, they were unable to do so because part was occupied by squatters. It was 19 days before the contractors could take possession of the whole of the site. The Court of Appeal held that the defendants were in clear breach of clause 21 because of their failure, for whatever reason, to give the plaintiffs possession. Although the contractors entered on the site, the trial judge found that they were unable to clear it and so the breach caused appreciable delay and disruption, which entitled the contractors to damages. This case should be contrasted with *Porter* v. *Tottenham Urban District Council* [1915], where the contractor was wrongfully excluded from the site by a third party for whom the employer was not responsible in law and over whom he had no control. There was no clause 21, and the court held that there was no implied warranty by the council against wrongful interference by a third party – an adjoining owner – with the only access to the site.

The phrase 'possession of the site' was considered in the 1985 *Whittal Builders* v. *Chester-Le-Street District Council* (the first Whittle case). It was held that the phrase meant possession of the whole site and that, in giving piecemeal possession, the employer was in breach of contract so as to entitle the contractor to damages. Mr Recorder Percival said:

'Taken literally the provisions as to the giving of possession must I think mean that unless it is qualified by some other words the obligation of the employer is to give possession of all the houses on 15 October 1973. Having regard to the nature of what was to be done that would not make very good sense, but if that is the plain meaning to be given to the words I must so construe them.'

Those words are a very clear statement of the law. Under JCT terms (both in 1963 and 1980 editions, before the 1987 amendment) there was no power for the architect to postpone the giving of possession of the site. This problem is less likely to arise under a JCT 98 contract, because clause 23.1 provides that if clause 23.1.2 is stated in the appendix to apply the employer may defer the giving of possession for a period not exceeding six weeks or such lesser period stated in the appendix.

An extension of time may be fixed under clause 25 and direct loss and/or expense may be claimed in respect of any deferment of possession by the employer. If the employer fails to give possession and the deferment provision is not stated to apply or if the failure lasts longer than stipulated by the provision, the employer will be in serious breach of contract as if the deferment provision had not been included.

7.9   *Time for completion*

The date for completion is defined in clause 1.3 as the date fixed and stated in the appendix. The date of possession is also defined as the date stated in the appendix, this time under clause 23.1.

The provisions in the contract regarding extensions of time are all to be found in clause 25 which is some three and a half pages long. The sole purpose of this clause is to provide the means of fixing a new date for completion. Whether or not the contract period is extended has no bearing whatsoever on the contractor's entitlement to loss and/or expense. Not only is it self-evident from the wording of the contract, it has been the subject of several judicial pro-nouncements, for example in *H. Fairweather & Co Ltd* v. *London Borough of Wandsworth* (1987). The fallacy that there is a connection has arisen because some of the grounds in the loss and/or expense clause (26) are echoed as grounds for extension of time under clause 25. In some quarters, it is prevalent for the quantity surveyor to include additional 'prelims' in a valuation after the architect has given an extension of time without any prompting from the con-tractor. Such a practice is quite wrong and borders on, if it does not actually overstep, negligence. Loss and/or expense, as distinct from additional preliminary items associated with a variation, can only be ascertained following an application in proper form from the contractor under clause 26 **[10.4]**.

7.10   *Relevant events*

Relevant events are listed in clause 25.4. The term refers to the grounds for which the architect is empowered to give an extension to the contract period. There are four possible categories of reason why the contractor may be delayed:

- Those which result from acts or omissions of the employer or the architect or others acting on behalf of the employer. This is the most important category, because if the architect was unable to give an extension of time for these reasons, the employer would forfeit his right to liquidated damages, e.g. by ordering variations as in *Dodd* v. *Churton* (1897). It is notorious that under this category, the architect is called upon to effectively sit in judgment on his own conduct. The relevant events in this category are: 25.4.5, 25.4.6, 25.4.8, 25.4.12, 25.4.13, 25.4.14, 25.4.17 and 25.4.18.
- Those where the contractor is delayed as a result of the acts or omissions of persons other than the employer or the contractor or those persons for whom one or other is responsible. The absence of these reasons from the list of relevant events would not jeopardise the employer's entitlement to liquidated damages. By including reasons from this category the employer is indicating that he is prepared to take the risk, at least of the delay potential. The relevant events in this category are: 25.4.4, 25.4.7, 25.4.9, 25.4.11 and 25.4.16.
- Those which amount to events which are outside the control of either of the contracting parties. The absence of these reasons from the list of relevant events would not jeopardise the employer's entitlement to liquidated damages. By including reasons from this category the employer is indicating that he is prepared to take the risk, at least of the delay potential. The relevant events in this category are: 25.4.1, 25.4.2, 25.4.3, 25.4.10 and 25.4.15.
- Those which are due to the contractor's actions or omissions. These are the risks which the contractor must take and, therefore, this category is entirely unrepresented in the list of relevant events except perhaps for clause 25.4.3 if the specified peril was the contractor's responsibility.

The relevant events will be dealt with in the same order.

7.11   Acts or omissions of the employer

7.11.1   Compliance with architect's instructions: clause 25.4.5

The instructions referred to in sub-section clause 25.4.5.1 are:

Clause 2.3   Discrepancies in contract bills, etc.
Clause 2.4.1   Discrepancies between documents.
Clause 13.2   Variations.

Clause 13.3   Expenditure of provisional sums in bills and sub-
                     contracts, but not if the expenditure is for defined
                     work or for performance specified work.
Clause 13A.4.1   Valuation of variations.
Clause 23.2   Postponement of any work to be executed under
                     the contract.
Clause 34   Antiquities.
Clause 35   Nominated sub-contractors.
Clause 36   Nominated suppliers.

The reference to clause 13.2 excludes a confirmed acceptance of a
clause 13A quotation. The reason for that is obviously because the
13A quotation and acceptance includes an adjustment to the time
required to complete the Works.

It should be noted that compliance with an architect's instruction
for the expenditure of a provisional sum for defined work or for
performance specified work is excluded. That is because in both
cases, the contractor has been given sufficient information to enable
him to make appropriate allowance in planning his work at tender
stage.

If there is a failure of a nominated sub-contractor, and there is
unreasonable delay by the architect in making a renomination, the
contractor is probably entitled to an extension of time under clause
25.4.5.1, with its specific reference to clause 35. It appears that the
question which arose in *Percy Bilton* regarding whether there was
power to extend time for delay in nomination is answered by this
clause [7.7].

Sub-clause 25.4.5.2 deals with opening up for inspection and
testing. The contractor may be entitled to an extension of time if
delay has been caused because the architect instructed that the work
was to be opened up for inspection or materials and goods to be
tested (clause 8.3), if the work inspected or the materials or goods
tested prove to be in accordance with the contract. Clause 8.4.4 has
elaborate provisions for repeated testing in the event of a defect
being shown in repetitive work. Here, again, an extension of time
may only be granted if the tests of other items of similar work show
that it is in accordance with the contract.

7.11.2   Instructions/drawings not in time: clause 25.4.6

This relevant event has been considerably slimmed down following
the introduction of the information release schedule [2.11] and the

substantially expanded wording of clause 5.4. The relevant event now falls into two parts. Where an information release schedule is used, clause 25.6.1 refers to the failure of the architect to comply with clause 5.4.1. What this means is that if the architect does not provide the information as set out in the schedule, the contractor has a ground for extension of time provided other criteria are met.

The second part of the relevant event is contained in clause 25.4.6.2. Not surprisingly, it refers to the failure of the architect to comply with clause 5.4.2 **[2.11]**. Clause 5.4.2 deals with the situation if an information release schedule has not been provided or if there is information required which is not listed on the schedule. Assessing delays under this ground is less easy than under clause 25.4.6.1, because there are no dates to act as reference points. It is much more a matter of judgement by the architect to decide when he was obliged to provide the information under clause 5.4.2 and whether or not he did so.

The terms to be applied about the time within which further drawings, details or instructions are to be given have been considered in relation to an engineer's obligations. It has been stated that such information must be given within a reasonable time, but it has been made clear that this is a limited duty:

> 'What is a reasonable time does not depend solely upon the convenience and financial interests of the [contractors]. No doubt it is in their interest to have every detail cut and dried on the day the contract is signed, but the contract does not contemplate that. It contemplates further drawings and details being provided, and the engineer is to have a time to provide them which is reasonable having regard to the point of view of him and his staff and the point of view of the employer as well as the point of view of the contractor.': *Neodox v. Borough of Swinton & Pendlebury* (1958).

This common-sense business approach is broadly applicable to the JCT wording. Under JCT contracts the architect does not control the order of the Works and, in clause 5.4.2, the phrase 'to enable the Contractor to carry out and complete the Works in accordance with the Conditions' must primarily be interpreted from the contractor's point of view. Factors to be taken into account will include the time necessary for the contractor to organise adequate supplies of labour, materials and plant and to execute any prefabrication or prepare materials in such time as to ensure that these things are available on site having regard to his obligation to complete the works in accordance with the contract: that is, by the contractual date for completion.

Mr Justice Vinelott said of similar provisions in JCT 63:

'What the parties contemplated by these provisions was first that the architect was not to be required to furnish instructions, drawings, etc. unreasonably far in advance from the date when the contractor would require them in order to carry out the work efficiently nor to be asked for them at a time which did not give him a reasonable opportunity to meet the request. It is true that the words "on a date" grammatically govern the date on which the application is made. But they are ... capable of being read as referring to the date on which the application is to be met. That construction seems to me to give effect to the purpose of the provision – merely to ensure that the architect is not troubled with applications too far in advance of the time when they will be actually needed by the contractor ... and to ensure that he is not left with insufficient time to prepare them. If that is right then there seems ... to be no reason why an application should not be made at the commencement of the work for all the instructions, etc. which the contractor can foresee will be required in the course of the works provided the date specified for delivery of each set of instructions meets these two requirements. Of course if he does so and the works do not progress strictly in accordance with this plan some modification may be required to the prescribed time-table and the subsequent furnishing of instructions and the like ... It does not follow that the programme was a sufficiently specified application made at an appropriate time in relation to every item of information required, more particularly in light of the delays and the rearrangement of the programme for the work': *London Borough of Merton* v. *Stanley Hugh Leach Ltd* (1985).

JCT 63 had a rather more substantial clause calling for a 'specific written application' to be made. Although that requirement is no longer present in the relevant event, there are clear echoes of it in clause 5.4.2.

7.11.3   Work not forming part of the contract: clause 25.4.8

Some years ago, clause 25.4.8 (then clause 23(h)) used to refer to 'artists and tradesmen'. It was familiarly known as the 'Epstein clause', because it appeared in the JCT contract after that sculptor had delayed the contractor by failing to produce his work on time and it was found that there was no provision to extend time for such an event.

Two separate situations are envisaged in this relevant event. The first situation is covered by clause 25.4.8.1 and deals with the situation where the employer engages other persons to carry out work under clause 29. The circumstances in which an extension of time can be given seem to be much broader than under the equivalent relevant event dealing with nominated sub-contractors (clause 25.4.7 **[7.12.2]**).

The second situation, dealing with the supply of materials which the employer has undertaken to supply, is found in clause 25.4.8.2. There are three points worth noting. First, in both situations it is not merely the employer's failure, but also the employer's success in correctly performing his obligations which may be grounds for extending time. Second, unlike the execution of work by others, there is no contractual provision which entitles the employer to provide materials or goods. The words which the employer 'has agreed' are not appropriate to a situation where supply of materials is a matter for the contractor and 'has elected with the agreement of the con-tractor' would better indicate that invariably any change to the supply by the contractor will be initiated by the employer. Third, an interesting scenario would be created if the materials subsequently were found to be defective after they had been built into the Works.

The decision in *Henry Boot* v. *Central Lancashire New Town Development Corporation* (1980) contains a useful explanation by Judge Fay of the meaning of the words 'not forming part of the contract'. Although concerned with the JCT 63 form of contract, the explanation is equally relevant to JCT 98.

Article 1 then, as now, provided that the contractor should exe-cute the work shown on the contract drawings and described or referred to in the contract bills, etc. and by article 2 that the employer would pay him the sum of £2,765,716 'hereinafter referred to as the "Contract Sum" '. The work by three statutory undertakers was contained in the contract bills under 'Direct Payments: Local Authorities and Public Undertakings – Provide the following sums for work to be executed ... Electric Main and Sub-station £28,500.'

There were similar provisions regarding water mains, gas mains and electrical connections to the street lighting system. There was also a further provision:

'The amounts included for Works to be executed by Local Authorities or public undertakings are to be expended under direct order of the central Lancashire Development Corporation.'

and

> 'Liaison with public bodies – The Employer intends to give per-
> mission for or instruct the following public bodies to carry out
> works during the progress of the Works: Local Authority;
> Highway Authority; Water Board; Gas Board; Electricity Board;
> Post Office. Afford all reasonable facilities to these bodies, and
> give ample notice when their work may proceed without inter-
> ruption and in accordance with the programme.'

The sums relating to work by these bodies were included in 'the
Contract Sum of £2,765,716'. Provision was made for the deduction
of the specified sums in the final account. The judge said:

> 'Just why this remarkable device has been adopted of putting this
> work in with one hand and taking it out with the other neither
> side in their hearing has satisfactorily explained . . .
> So now I reach the position that by Article 1 of the contract the
> contractor binds himself to carry out works which, under the bills
> of quantities, incorporated into the contract, he is told not to do,
> and he is told moreover that others will do.
> And under Article 2 he is to be paid a total sum, including part
> for work which he is not to do and which the bills of quantities
> provide that he shall not receive.
> Does or does not this work form part of the contract?'

The judge considered the implications of what is now JCT 98 clause
2.2.1 to the effect that nothing in the bills should override or modify
the printed articles, conditions or appendix and decided, following
English Industrial Estates v. *Wimpey* (1973), that he was entitled to
look at the bills to 'follow what was going on'. He concluded:

> 'For some purposes the work does form a part, literally, of the
> contract; but for other purposes it does not.
> It is not work which the employer can require the contractor to
> do. All that he can require is that the contractor affords atten-
> dance etc. on those who do the work . . . [and that] I take the
> pragmatic view that the relevant work is work not forming part of
> the contract.'

7.11.4   Failure to give ingress and egress: clause 25.4.12

There is no right in the contract (or power in the architect) to grant
an extension of time under this sub-clause for the employer's failure

to give possession of the site itself on the date for possession. This is of particular importance where the employer has not taken power to defer the giving of possession of the site to the contractor or where the deferment exceeds the period allowed in the contract.

An extension of time can only be granted under this sub-clause where there is failure by the employer to provide access to or exit from the site of the works across any *adjoining or connected* 'land, buildings, way or passage' which is in his own *possession and control*. It does not cover, for instance, failure to obtain a wayleave across an adjoining owner's property, or where, for example, access to the highway is obstructed because the local authority has temporarily closed the road: *National Carriers Ltd* v. *Panalpina (Northern) Ltd* (1981). Such a temporary closure might amount to frustration of the contract if it lasted long enough. It would equally not extend to the situation where protestors or perhaps strikers impeded access to a site where contractors were carrying out work: *LRE Engineering Services Ltd* v. *Otto Simon Carves Ltd* (1981).

The wording in the sub-clause itself refers to access, etc. 'in accordance with Contract Bills and/or Contract Drawings', which suggests that the undertaking to provide such access must be stated in the bills or drawings, and in that case any extension of time will be dependent upon the contractor giving whatever notice may be required by the provision in the bill before access is to be granted. But the clause goes further with its reference to 'or failure of the Employer to give such ingress or egress as otherwise agreed between the Architect and the Contractor'. This would seem to give the architect the authority to reach such an agreement as agent on the employer's behalf so that the employer in effect becomes responsible for failing to honour such an agreement even though it may have been reached without his being consulted. This, if correct, seems an extraordinary extension of the architect's powers. The architect's authority as the employer's agent is a limited one in law and would not normally extend this far.

It is to be noted that the contract imposes no strict liability on the employer to ensure access to the contractor.

7.11.5   Deferment of possession: clause 25.4.13

Failure by the employer to give possession of the site is quite common. Clause 23.1 enables the employer to defer giving the contractor possession of the site for a period of up to six weeks unless he has inserted a shorter period in the appendix. He only has this power if it is expressly stated in the appendix that clause 23.1.2 applies. Where the employer does defer the giving of possession, there will be entitlement to extension of time. In practice, that often has the simple effect of moving the contract period bodily backwards so that the extension of time equals the period by which the date of possession is deferred. It may not always be quite so simple, because the contractor, having been deprived of the expected date of possession, may need an additional time in which to organise for the new date of possession. Everything will depend upon the particular facts.

7.11.6 Approximate quantity not a reasonably accurate forecast of the quantity of work: clause 25.4.14

This ground is set out on the perfectly reasonable basis that a contractor will plan his work using, among other things, the quantities in the bills of quantities. Where such quantities are described as 'approximate', it is presumably because the architect and/or the quantity surveyor either does not know, or has not quite decided upon, the amount required. All the contractor can do is to use the approximate quantities as if they gave a reasonably accurate forecast of the quantities required. Indeed the contractor, in preparing his programme, can do no other than assume that the quantities are strictly accurate. The author once encountered an architect who was of the opinion that the contractor should allow a margin of 20% either way for all approximate quantities. How he was to do that was never made clear. If the quantities give a lower forecast, he is entitled to an extension to represent the additional time required to carry out the work.

7.11.7 Compliance or not with clause 6A.1: clause 25.4.17

The employer's obligation under clause 6A.1 is to ensure that the planning supervisor carries out his duties under the CDM Regulations 1994 and, if the contractor unusually is not the principal contractor under the regulations, to ensure that he carries out his duties also. It should be noted that the ground encompasses 'compliance or non-compliance' so that the proper carrying out of duties may also attract an extension of time. The employer's obligation to 'ensure' is an onerous one. The planning supervisor has duties under the regulations which he may have to carry out after the issue of any architect's instruction. Therefore, each instruction may attract an extension of time under this ground even if it does not qualify under clause 25.4.5. Although anecdotal evidence suggests that this avenue has not yet been thoroughly explored by contractors, the possibilities for extensions of time are almost limitless.

7.11.8   Delay arising from clause 30.1.4 suspension: clause 25.4.18

Clause 25.4.18 introduces this ground to comply with section 112 of the Housing Grants, Construction and Regeneration Act 1996 which entitles a contractor to suspend performance of his obligations on seven days written notice if the employer does not pay a sum due 'in full by the final date for payment'. The suspension part of section 112 is dealt with by clause 30.1.4 **[9.15]**. This relevant event covers section 112(4) which states:

> '(4) Any period during which performance is suspended in pursuance of the right conferred by this section shall be disregarded in computing for the purposes of any contractual time limit the time taken, by the party exercising the right or by a third party, to complete any work directly or indirectly affected by the exercise of the right.'

No one could honestly say that the draftsman had produced a paragraph of crystalline clarity. However, it appears that the Act provides that if a party suspends performance for six days, the effective extension to the period for completing the work is to be six days. This ignores any time the contractor may need to get ready to recommence. The wording of clause 25.4.18, by referring to 'delay arising from a suspension', clearly requires the architect to consider all the delay and not just the actual period of suspension.

7.12   *Acts or omissions of others*

7.12.1   Strikes and similar events: clause 25.4.4

'Civil commotion' means, for insurance purposes, 'a stage between a riot and a civil war': *Levy* v. *Assicurazioni Generali* (1940). So far as strikes are concerned, it should be noted that extension may be given, not only for circumstances affecting the contractor himself and his work on site, but also those engaged in preparing or transporting any goods and materials required for the works. The wording covers both official and unofficial strikes, but it does not cover 'working to rule' or other obstructive practices which fall short of a strike. It has been held that a strike by workers employed by statutory undertakers directly engaged by the employer to execute work not forming part of the works was not covered by the forerunner of this clause in JCT 63: *Boskalis Westminster Construction Ltd* v. *Liverpool City Council* (1983). A strike or other event referred to in the sub-clause must be one in which the trades mentioned in it are directly involved.

The expression 'local combination of workmen' is an old phrase of imprecise meaning, but apparently beloved of the draughtsmen of insurance policies. It might be held to cover obstructive activities falling short of a strike provided they were confined to one area or site.

7.12.2 Delay by nominated sub-contractors and suppliers: clause 25.4.7

The words 'delay on the part of' are repeated in clause 25.4.7 in spite of numerous judicial criticisms and notwithstanding Lord Wilberforce's observation in *Westminster Corporation* v. *J. Jarvis & Sons* (1970): 'I cannot believe that the professional body, realising how defective this clause is, will allow it to remain in its present form'. On the other hand, the JCT can now argue, with some justification, that there is every reason to retain the clause in its original form, because we all have the benefit of the decision of the House of Lords on its true meaning.

The words do not mean delay *caused* by a nominated sub-contractor or supplier. Nor does delay mean just sloth or dilatoriness. Again, Lord Wilberforce said in the same case: '. . . it is contractually irrelevant whether a sub-contractor could have worked faster'. It means solely and exclusively failure to complete the sub-contract works by the contractual date. It does not include delay caused to the contractor by the repudiation or insolvency of the sub-contractor. As Lord Wilberforce said: 'If it were, why should the word "delay" be used? Why not frankly exonerate the contractor for any delay in completion due to any breach of contract or failure, *eo nomine* of the sub-contractor.' In other words, if the draftsmen had intended that the contractor should get an extension of time for any delay *caused by the sub-contractor*, they would have used those words.

If a nominated sub-contractor (or, for that matter, a nominated supplier) ostensibly completes his sub-contract work or his supply contract but later is found to be in breach, e.g. because defects appear in the work, and has to return to remedy the breach, that is not a 'delay' within the meaning of this sub-clause.

The delay referred to by the sub-clause must, it is noted, be delay which the contractor has taken all practicable steps to avoid or reduce, and perhaps the observations of Viscount Dilhorne in *North West Metropolitan Regional Hospital Board* v. *Bickerton* as to the general legal situation will serve as a warning to main contractors. He said:

> 'I cannot myself see that the extent of the contractor's obligation ... is in any respect limited or affected by the right of the architect to nominate the sub-contractors. He has accepted responsibility for the carrying out and completion of all the contract works including those to be carried out by the nominated sub-contractor. Once the sub-contractor has been nominated and entered into the sub-contract, the contractor is as responsible for his work as he is for the works of other sub-contractors employed by him with the leave of the architect.'

However, it is clear that 'delay by the employer in making the timeous nomination of a new sub-contractor is within the express terms of clause 23(f)' of JCT 63 and, of course, within JCT 98, clause 25.4.6: *Percy Bilton Ltd* v. *Greater London Council* (1982). It is for the contractor to make application to the architect for a renomination in respect of failure by a nominated sub-contractor or nominated supplier.

The basic scheme of nomination aims to give the employer who has lost the right to liquidated damages against the contractor, due to some fault of the nominated sub-contractor, recourse in damages against the nominated sub-contractor concerned **[7.7]**.

7.12.3   Government action: clause 25.4.9

The government action must take place after the 'Base Date' to qualify as a ground under this relevant event. The 'Base Date' is a date written into the appendix. In the case of JCT 80 before its amendment 11 July 1987, the reference was to the 'Date of Tender', meaning '10 days before the date fixed for receipt of tenders by the employer' (clauses 38.6.1 and 39.7.1 in their original form), which did not always in practice provide a firm date due to the frequency with which the date for receipt of tenders was extended.

This provision might, for example, be relied on in a 'three-day-week' situation or wherever the British Government exercises any statutory power in the sense described in the clause, e.g. under the Defence of the Realm Regulations. The real significance of this is that it is no longer to be covered by *force majeure,* and the contract would be prevented from being brought to an end by frustration. A prolonged stoppage of work for this reason would not, therefore, be grounds for the contractor to determine his own employment under clause 28. That can only be good news for the employer.

7.12.4   Work by statutory undertakers or a local authority: clause 25.4.11

These provisions cover delay caused by 'the carrying out by a local authority or statutory undertaker of work *in pursuance of its statutory obligations* in relation to the Works, or the failure to carry out such work'. In *Henry Boot Construction Ltd* v. *Central Lancashire Development Corporation* (1980), which arose from an award made in the form of a special case by an arbitrator, the judge was concerned with the problem of whether 'statutory undertakers' were 'artists, tradesmen or others engaged by the Employer' for the purpose of JCT 63, clauses 23(h) and 24(1)(d). They were 'engaged by the employer in carrying out work not forming part of this contract'.

This decision has made no difference whatever to the perfectly clear meaning to be attached to these words, which the arbitrator obviously had very firmly in mind. Extensions of time can only be granted under this head if the statutory undertakers are carrying out work that is a statutory obligation. Statutory undertakers frequently do work that is not done under statutory obligation, even though only they can do it. In such a case, if they have been 'engaged by the employer', any extension of time would be made under clause 25.4.8.1, with a possible claim for direct loss and/or expense under clause 26.2.4.1. If, however, they were engaged by the contractor, it seems that no extension of time would be due.

Statutory undertakers may, of course, affect the work in other ways. For example, a water authority might be laying water mains in the public road which provides access to the site, not for the purposes of the particular contract works but perhaps for another site nearby. In that case, even though they might be under an obligation to lay the mains (and be carrying out the work as a matter of statutory obligation) they would not be doing so in relation to the works. In such a case, there could be no extension of time under clause 25.4.11 because the statutory undertaker would not be carrying out 'work ... in relation to the Works' and, indeed, it seems to us that there is no provision in JCT 80 under which an extension of time could be given. (Dubiously, it might be argued that such activities constituted *force majeure*.)

In practical terms, whether or under which sub-clause an extension of time should be given for delays caused by local authorities and statutory undertakers depends entirely upon the sort of work they are doing and the circumstances under which they are doing it.

7.12.5   Use or threat of terrorism: clause 25.4.16

It is thought that the threat would have to be more substantial than just the fact that other terrorist incidents have occurred in the area. A specific terrorist threat directed at the project or a threat to an area which, if it was to be carried out, would affect the project would qualify. The activities of the relevant authorities which would qualify under this ground would include such measures as evacuation of premises and the restriction of access. This ground is not restricted to the site of the works and, therefore, it is likely that any such threat or action which affected the execution of the works in any way (such as the forced evacuation or destruction of the contractor's offices) would give entitlement to extension of time.

7.13   *Events outside the control of either party*

7.13.1   *Force majeure*: clause 25.4.1

Force majeure is a French law term which is wider in its meaning than the common law term 'Act of God'. Under JCT contracts the term *force majeure* has a restricted meaning because many matters such as war, strikes, fire and exceptional weather are expressly dealt with later in the contract. There appear to be no reported cases dealing with the matter in the context of JCT contracts, and the authority usually quoted is *Lebeaupin* v. *Crispin* (1920), where Mr Justice McCardie accepted that:

'This term is used with reference to all circumstances independent of the will of man, and which it is not in his power to control... Thus war, inundations and epidemics are cases of *force*

majeure; it has even been decided that a strike of workmen constitutes a case of *force majeure* . . . [But] a *force majeure* clause should
be construed in each case with a close attention to the words
which precede or follow it and with due regard to the nature and
general terms of the contract. The effect of the clause may vary
with each instrument.'

Decisions on the meaning of the word when used in other forms of
contract are of little assistance. The dislocation of business caused
by the general coal strike of 1912 has been held to be covered by the
term and also covered the breakdown of machinery, but not delay
caused by bad weather, football matches or a funeral: *Matsoukis* v.
Priestman & Co (1915).

The event relied upon must make the performance of the contract
wholly impossible and, in this sense, the term is similar to the
English law doctrine of frustration of contract.

7.13.2   Exceptionally adverse weather conditions: clause 25.4.2

The change in wording in the 1980 form from 'inclement' to
'adverse' was intended to make it clear that the ground is intended
to cover any kind of adverse conditions including drought. The
emphasis is on the word *exceptionally* and the meaning of the phrase
is to be found by considering two things. First, the kind of weather
that may be expected at the site at the particular time when the
delay occurs. Second, the stage which the works have reached.

In regard to the first factor reference to local weather records may
be helpful in showing that the adverse weather was 'exceptional' for
that area, i.e. exceeding what may on the evidence of past years be
reasonably expected. Reference to at least the previous five years
records would be required. The dictionary meaning of adverse is
'contrary' or 'hostile'. 'Exceptionally' means 'unusual'. In regard to
the second factor even if the weather is exceptionally adverse for the
time of year it must be such that it interferes with the works at the
particular stage they have reached when the exceptionally adverse
weather occurs, even though the Works may be affected entirely
due to the contractor's own fault; for example, where he failed to get
the roof on in accordance with his own programme and exceptionally heavy rain makes it impossible to continue work: *Walter
Lawrence* v. *Commercial Union Properties* (1984). If despite the
weather, works could continue then the works have not been
delayed by the exceptionally adverse weather. The contractor is

expected to programme the works making due allowance for normal adverse weather, i.e. the sort of weather which is to be expected in the area and at the time of year during the course of the works. His programme for those parts of the work which may be affected by rain, wind or frost should acknowledge the fact that interruption is likely to occur, and should allow for it.

7.13.3   Specified perils loss or damage: clause 25.4.3

This is intended to give the contractor the necessary time to fulfil his obligations to repair damage caused by fire, lightning, explosion, storm, tempest, flood, bursting or overflowing of water tanks, apparatus or pipes, earthquake, aircraft or other aerial devices, or articles dropped therefrom, riot and civil commotion, but excluding what are called the 'Excepted Risks': These are defined in clause 1.3.

The only important practical question arising under this heading is whether or not the contractor is entitled to an extension of time where the events are caused by the default or negligence of the contractor's own employees. On the plain reading of the wording it would appear that the test is simply whether the delay is caused by one of the specified perils. The contractor's negligence, if it exists, is irrelevant to that question although it is relevant to the related question of whether an occurrence falls within the category of specified perils **[8.10]**.

7.13.4   Failure to obtain labour and goods: clause 25.4.10

The date at which any shortage is to be unforeseeable is the 'Base Date'. There are two sub-clauses. One deals with labour (clause 25.4.10.1), the other deals with materials (clause 25.4.10.2). In order to qualify as a relevant event, not only must the shortage be reasonably unforeseeable, the inability to obtain labour or materials must be for reasons which are beyond the contractor's control. Although there will be certain fairly rare instances when the contractor will not be able to obtain certain materials no matter what measures he takes or what price he is prepared to pay, he will always be able to obtain labour. Sometimes he will have to pay a grossly inflated price, but he will always be able to secure labour. In order to make sense of this particular event it is necessary to make some implication regarding the availability of labour or materials at prices which could reasonably be assumed by the parties at the base date. This is a peculiarly difficult event to consider in practice, because the contractor will always contend that the price which he put in his tender was the critical price above which labour had to be considered unobtainable. In some forms, but not JCT 98, it is optional.

7.13.5 Alteration to performance specified work due to change in statutory requirements: clause 25.4.15

This relevant event applies to the situation when, after the base date stated in the appendix, there is a change in statutory requirements which makes necessary an alteration in performance specified work to be carried out by the contractor under clause 42. Such a change might be an amendment to the Building Regulations which obliges the contractor to revise his proposals. It seems that there would be two possible bases for extension: if the redesign involves delay or if the changed or additional work takes longer to execute. The contractor must have taken all practicable steps to avoid or reduce the delay and, of course, there are no grounds for extension of time if the change occurred before the base date and, therefore, could have been taken into account as part of the contractor's tender [3.13].

7.14   Best endeavours

Clause 25.3.4 introduces two very important provisos which the architect is obliged to consider before giving any extension of time in any circumstances. Under clause 25.3.4.1, the contractor must use 'constantly his best endeavours to prevent delay'. This seems to be no more than an express restatement of the contractor's common law obligation and reinforcement of the express provisions of the contract relating to completion and the contractor's obligation to proceed regularly and diligently.

It has been said by some commentators that 'best' means 'best'; in other words – everything within the contractor's power, using any means. There is some justification for this point of view, particularly in light of the use of the phrase obligating the contractor to 'do all that may reasonably be required' in the second part of the proviso, which is in contrast with the 'best endeavours' obligation. There appears to be no construction industry case directly in point, but in other contexts using best endeavours has been held to mean doing everything prudent and reasonable to achieve an objective: *Victor*

Stanley Hawkins v. *Pender Bros Pty Ltd* (1994). Clearly, it is a lesser obligation than to 'ensure' or to 'secure', which impart an absolute liability to perform the duty set out: *John Mowlem & Co* v. *Eagle Star & Others* (1995). If the architect's decision is that the contractor has not *constantly* used his best endeavours, then the contractor's only recourse is to challenge that decision in adjudication or arbitration.

The second proviso is to be found in clause 25.3.4.2. It states that the contractor must do everything 'that may reasonably be required to the satisfaction of the Architect to proceed with the Works'. This is the contractor's obligation in any case, but the architect has no power to order that acceleration measures be taken (either under this provision or any other provision in the contract). If 'best endeavours' actually obliged the contractor to do everything in his power including accelerating the progress, the extension of time clause would be redundant, because the contractor's obligation would always be to accelerate sufficiently to avoid delays no matter what their origin.

The two provisos are complementary. The first requires the contractor to show initiative, perhaps by reprogramming the Works, the second requires him to take account of any requirement of the architect. For example, the architect may have his own firm views on the kind of reprogramming which would be most effective. Leaving aside the wisdom of the architect meddling in such matters which are essentially the province of the contractor, the contractor must comply with the architect's requirement provided that it is *reasonable*. It can hardly be considered reasonable for an architect to require a contractor to do something which will cost the contractor significant amounts, or even any amount, of money.

7.15 Contractor's notice of delay

It should be noted that, under clause 25.2.1.1, the contractor is to give notice not only when the progress of the works is being delayed, but also when it becomes reasonably apparent that it is likely to be delayed in the future. It has to be reasonably apparent that the progress of the works is being or is likely to be delayed. It is the actual progress and not the contractor's planned progress which is relevant, as the wording makes clear: *Glenlion Construction Ltd v. The Guinness Trust* (1987). Apparent means 'manifest', presumably to the contractor, and once it becomes reasonably apparent that the progress is actually being delayed or is likely to be delayed, the contractor must notify the architect in writing. The notice must specify the cause of delay. Under JCT 98 wording, the architect has power to give an extension in the absence of written notice; the contractor's failure to give written notice surely means that the architect does not need to make a decision on extensions until his review of the completion date not later than the expiry of 12 weeks from the date of practical completion, because clause 25.3.3.1 expressly states: 'whether or not the Relevant Event has been specifically notified by the Contractor'. It is thought that he is not entitled to give an extension of time before practical completion in the absence of written notice.

The contractor's notice is to state not just the 'cause or causes' of the delay; it must also state 'the material circumstances'. It should go into some detail as to why and how the delay is occurring or is likely to occur; the 'material circumstances' will include, for example, the stage the contract has reached, the proposed order of works, and so on. The duty is not limited to notifying the causes of delay listed as relevant events; it is a duty to give notice of delay for any reason. This is so even if it is uncertain whether the current completion date will be affected. The purpose of the notice is simply to warn the architect of the situation, and it is then up to him to monitor it. He may, if necessary, take remedial action and forewarn the employer. The contractor must give prior notice of delays which it is reasonable for him to anticipate. This gives the architect the opportunity, once he has been notified of any impending delay, to take steps to rectify the situation and bring the contract back on schedule. One of the things the architect can do is to omit work under clause 13.2. If the contractor fails to give notice of a delay which he clearly should have been able to anticipate, the architect can say that the contractor has not used his best endeavours to prevent delay in progress, which he is bound to do under the terms of clause 25.3.4.1. If the contractor fails to give notice, he is in breach of contract and the architect is entitled, and probably obliged, to take the breach into account when considering a future extension of time. The notice must state those causes of delay which, in the contractor's opinion, entitle him to an extension of time. The 'Relevant Event' so identified must be one (or more) of those listed in clause 25.4 and it is for the contractor to specify them.

Clause 25.2.1.2 introduces a further requirement, and that is that a copy of the contractor's original notice must be sent to any nominated sub-contractor to whom reference is made in it. One of the 'Relevant Events' listed is 'delay on the part of nominated sub-contractors or nominated suppliers which the contractor has taken all practicable steps to avoid or reduce' (clause 25.4.7). The purpose of giving a copy of the notice to affected nominated sub-contractors is to forewarn the nominated sub-contractor so that he may in turn, if necessary, make application for extension of time to the main contractor under clause 2.2 of the Nominated Sub-Contract Form NSC/C. Clause 25.2.2 imposes an additional obligation on the contractor. Either in his original notice or, where that is not practicable, as soon as possible after the notice, the contractor must state in writing to the architect *particulars of the expected effects* of each and every relevant event identified in the notice, i.e. particulars of the expected effects on progress, and each and every relevant event must, for this purpose, be considered in isolation. The contractor must provide sufficient detail to enable the architect to make a proper decision.

The contractor must give his own estimate of the expected delay in completion of the works beyond the completion date 'whether or not concurrently with delay resulting from any other Relevant Event'. This is a particularly onerous task rarely undertaken correctly. The contractor must address each delay and its effects separately even if two or more delays are acting together. The particulars and estimate must be 'reasonably sufficient' to enable the architect to form a judgment (clause 25.3.1). A copy of the contractor's particulars and estimate must be given to any nominated sub-contractor to whom a copy of the original notice was given under clause 25.2.1.2.

By the terms of clause 25.2.3, each notice of delay must be kept under review by the contractor and he must revise his statement of particulars and estimate and/or give whatever further notices *may reasonably be necessary or as the architect may reasonably require.* The duty extends to *any material change* in the particulars, etc. The contractor must keep the architect up to date with developments as they occur and, importantly, the contractor's duty is not dependent upon the architect's request. The architect's time limit for dealing with applications for delay only starts to run when he has received 'reasonably sufficient particulars and estimates' from the contractor (clause 25.3.1). The clear intention of these provisions is to provide the architect with sufficient information reasonably to form his own judgment on the matter. He may not have been on the site at the time of the delay, though he must use whatever records he has: *London Borough of Merton* v. *Stanley Hugh Leach Ltd* (1985). Affected nominated sub-contractors must also be kept informed.

7.16   Architect's response

It is entirely a matter for the architect to decide whether, in his opinion, a delay in the contract completion date is likely to occur or has occurred and also whether the cause of delay is one of those listed in clause 25 and therefore one for which he should grant an extension. Plainly, in making his decision, the architect is obliged to follow any guidelines established by law.

Once the architect is notified by the contractor of delay, it is for him to monitor the position. The position was aptly stated by Mr Justice Vinelot in *London Borough of Merton* v. *Stanley Hugh Leach Ltd* (1985):

> 'The architect is entitled to rely on the contractor to play his part by giving notice when it has become apparent to him that the progress of the works is delayed. If the contractor fails to give notice forthwith upon it becoming so apparent he is in breach of contract and that breach can be taken into account by the architect in deciding whether he should be given an extension of time. But the architect is not relieved of his duty by the failure of the contractor to give notice or to give notice promptly. He must consider independently in the light of his knowledge of the contractor's programme and the progress of the works and of his knowledge of other matters affecting or likely to affect the progress of the works ... whether completion is likely to be delayed by any of the stated causes. If necessary he must make his own inquiries, whether from the contractor or others.'

That was said of the extension of time provision in JCT 63. It is probable that under JCT 98, the architect may not make an extension of time without the contractor's notice until after the contract completion date (or any extension to it) has passed.

If the contractor feels that the architect has been unreasonable in reaching his opinion, his recourse is to adjudication or to arbitration. On receipt of the contractor's written notice the architect must decide if the cause of delay is covered by clause 25. If in his view it is not, then subject to the contractor's right to challenge that opinion by one of the dispute resolution methods, that is the end of the matter.

The architect must not arrive at his decision on a whim. He should carefully analyse the position and consider the effect of individual delays: *John Barker Construction Ltd* v. *Portman Hotels Ltd* (1996). It is only when he has received the notice, particulars and estimate from the contractor that the architect must consider them before the date for completion. The architect must then decide: (1) Whether any of the causes of delay specified by the contractor in the notice is in fact a relevant event. He may disagree that the particular cause specified by the contractor is a relevant event, in which case the architect need not consider the next point. (2) Whether completion of the works is in fact likely to be delayed *thereby* (i.e. by the specified relevant event) beyond the completion date. Then, and only then, does his duty to give an extension of time arise. The architect must then decide whether or not the delay is going to mean a likely failure to complete by the date for completion. In making up his mind on this point the architect is entitled to consider the proviso to the clause that the contractor shall constantly use his best endeavours to prevent delay. The contractor's duty is to prevent delay, so far as he can reasonably do so. A delay in progress of the works at an early stage may be reduced or even eliminated by the contractor using his best endeavours.

It should be noted that the obligation to make an extension appears to rest on the architect without the necessity of any formal request for it by the contractor. He is required to do this only if the completion of the works 'is likely to be or has been delayed beyond the Date for Completion', or any extended time for completion previously fixed. Under clause 23.1.1, the contractor is under a double obligation: on being given possession of the site, he must begin the works and regularly and diligently proceed with them, and he must also complete the works on or before the date for completion, subject to any extension of time. If a strike occurs when two-thirds of the work has been completed in half the contract time, on resuming work a few weeks later the contractor is not then entitled to slow down the work so as to last out the time until the date for completion (or beyond, if an extension of time is granted) if as a result he is failing to proceed with the work 'regularly and diligently': *London Borough of Hounslow* v. *Twickenham Garden Developments Ltd* (1970). The consequence is that the architect is entitled to take into account the fact that the contractor is in advance of programme when considering what extension to grant; and he may also make use of the contractor's 'float' element in the contract programme: *How Engineering Services Ltd* v. *Lindner Ceiling Partitions plc* (1995). Where there are overlapping causes of delay, the architect must consider each cause separately, so that if there is one ground justifying an extension and another not, the architect cannot deprive the contractor of any reasonable extension for the relevant event merely because there is an overlapping cause; but the cumulative effect on progress must be taken into account: it is delay to progress which is the important factor. The use of networks, programmed on the computer, is very useful in these circumstances.

The architect must give an extension of time to the contractor if in his opinion there is a relevant event and that relevant event is likely to delay the completion of the works. However, in deciding what extension of time to give, if any, the architect will take into account any overlapping of delays resulting from different relevant events and, presumably, he will do so after study of the contractor's master programme if this has, in fact, been provided under clause 5.3.1.2. Where there is such a master programme, suitably annotated, it should be one of the 'reasonably sufficient particulars' to be provided by the contractor in support of his original notice. The architect's obligation is to grant an extension of time by fixing such later date as the completion date as he then estimates to be fair and reasonable, and where the contractor is in advance of planned progress, or does not at that time actually need the extension, the architect is not bound to so grant it: *London Borough of Hounslow* v. *Twickenham Garden Developments Ltd* (1970). Having received from the contractor all the requisite information, the architect's duty is to consider the information provided and to make his own assessment of the situation as to whether or not an extension should be granted at that time and, if so, what extension should be granted.

There is no contractual obligation to provide a master programme. It is common and sensible practice for the architect to include such a requirement, and specify the type of programme required, in the bills or specification, in which case the requirement will become a contractual provision: *Glenlion Construction Co Ltd* v. *The Guinness Trust* (1987). As a minimum, the architect should require a programme in bar chart *and* in network form with key dates and resources clearly shown. The architect is required to grant the extension of time by fixing as a new completion date for the works *'such later date ... as he then estimates to be fair and reasonable'*. The architect is only expected to estimate the length of extension and not to ascertain. Ascertainment, in the sense of 'finding out with certainty' would be impossible.

The architect must inform the contractor in writing of the new completion date, and he must state two things: which of the relevant events he has taken into account, and the extent, if any, to which he has had regard to any omission instruction issued since the fixing of the previous completion date. If the architect has not issued any omission instruction, the only information he must give to the contractor is the new completion date and the relevant events taken into account. There is no obligation for the architect to state the period of time he has allocated to each event. The contractor will, of course, demand these details, because without them it is very difficult to challenge the architect's decision unless it is grossly wrong. The architect would be obliged to reveal his calculations during an adjudication or arbitration if he wanted to defend his decision. The architect may take account of omissions when he decides on the first extension application that he grants because the definition of completion date in clause 1.3 is 'the Date for Completion as fixed and stated in the Appendix or any later date fixed' under the relevant provisions, and date for completion includes the original date. It follows that there is always a 'previous completion date' for the purposes of clause 25.3. The only proviso is that the architect cannot fix a completion date earlier than the date for completion stated originally in the appendix – something which is forbidden by clause 25.3.6. If he takes account of omissions, the architect must inform the contractor in writing when fixing the new completion date.

There is a time limit of 12 weeks from receipt of the contractor's notice of delay and of 'reasonably sufficient particulars and estimate' from the contractor, in which the architect must reach a decision and, if he considers it appropriate, give an extension of time. He is only required to comply with this time limit if it is reasonably practicable to do so. The correct operation of these provisions really depends on both architect and contractor being of one mind as to whether the information supplied by the contractor is 'reasonably sufficient' to enable the architect to reach a decision. From the employer's point of view it is important that the architect should decide quickly because of the fluctuations provisions, the effect of which is that, unless the architect carries out his duties timeously, the right of the employer to freeze the contractor's fluctuations on the due date for completion is lost **[5.18]**.

If there are fewer than 12 weeks left between receipt of the contractor's notice, particulars and estimate and the currently fixed completion date, the architect must reach his decision and grant any extension no later than that date. The intention is that the contractor should always have a date before him. If there is a very short period left before the completion date, it may not be 'reasonably practicable' for the architect to come to a decision in time. In such a case, his duty is probably to make the best decision practicable, which may well be a conservative one, before the completion date and then use the additional period thus created to come to a more considered decision. Thus, an architect faced with making a decision just one week before completion date may decide he is able to give two weeks extension of time and the extra two weeks may enable him, on mature reflection, to give a further one week. In some instances, the architect may not find it reasonably practicable to make any decision before the completion date and, in consequence, he may leave it for the review under clause 25.3.3. Some architects have adopted the practice of amending clause 25 so as to do away with the time limits. This is, in our view, a most unwise practice, if only because of the fluctuations provisions. Fluctuations are only to be frozen at completion date 'if the printed text of clause 35 is unamended and forms part of the Conditions': see clause 38.4.8; 39.5.8; and 40.7.2.

The final paragraph of clause 25.3.1 is important and its effect is that in respect of each notification of delay under clause 25.2 after the provision of any further particulars and estimates required, the architect must notify the contractor in writing if his decision is not to fix a later completion date as a new completion date. It is important, because the architect's decisions are required before the provisions restricting the level of fluctuations or formula can be operated if the contractor is in default over completion **[5.18]**.

Clause 25.3.2 is much misunderstood. After the first extension of time that the architect gives or after a revision to the completion date stated in a confirmed acceptance of a clause 13A quotation **[6.10]**, the architect can use his powers under the clause. He cannot, in any case, fix any earlier date than the original completion date: clause 25.3.6. But if he has issued instructions which result in the omission of work or obligations under clause 13.2 or under clause 13.3 in regard to provisional sums for defined work or for performance specified work, he is entitled to take this into account and 'fix a Completion Date earlier than that previously fixed under clause 25 if in his opinion the fixing of such earlier completion date is fair and reasonable' having regard to those instructions. Each extension is deemed to take into account all omissions instructed up to the date of the extension.

If architects wish to take advantage of the power to reduce extensions previously granted on account of omissions of work or obligations, they should take the decision and notify the contractor at the earliest possible moment – preferably when issuing the instruction – and not leave it until they next give an extension of time. The reason for this is that architects may be conservative in giving extensions of time, knowing that they can, at the end of the day, grant a little more time. To err too much in the direction of parsimony – and to be unrealistic in considering the effect of omissions – is not good practice.

7.17   *The architect's review*

Clause 25.3.3 gives the architect the opportunity to make a final decision on extensions of time. This clause imposes a mandatory obligation on the architect to review the completion date in any event not later than 12 weeks from the date of practical completion. In *Temloc Ltd* v. *Errill Properties Ltd* the Court of Appeal appeared to hold that the requirement is not mandatory. This is a wrong view of the judgment **[7.6]**.

Clause 25.3.3 requires the architect to review the completion date in any event; and he must do this in light of any relevant events whether or not specifically notified to him by the contractor. The opening sentence makes it clear that the architect must take account of any relevant events which occur during the period between the contract completion date and practical completion, which is intended to be the architect's final opportunity to consider extensions of time. It is at least arguable, on a strict reading of clause 25.3.3, that the architect can exercise this power only once. Therefore, if he chooses to do so before practical completion, it may be that he cannot do it again afterwards. In practice, an architect will usually wait until after practical completion to act under this clause. The architect has no discretion; he must write to the contractor and do one of three things:

(1)   *Fix a completion date later than that previously fixed:* He must do this if in his opinion so to do is 'fair and reasonable having regard to any of the Relevant Events', i.e. those listed in clause 25.4, *'whether upon reviewing a previous decision or otherwise'* and 'whether or not the Relevant Event has been specifically notified by the Contractor'.
(2)   *Fix a completion date earlier than that previously fixed:* He must do this if in his opinion it is fair and reasonable to do so having regard to any omission instructions issued, in connection with variations, provisional sums for defined work or performance specified work, since he last granted an extension of time.
(3)   *Confirm to the contractor the completion date previously fixed.*

7.18 *Variations after completion date*

A perennial question concerns the position if the architect gives an instruction requiring a variation after the date for completion and during a period when the contractor is in culpable delay. The question is really a number of questions:

- Is the architect entitled to instruct a variation after the date for completion has passed?
- If the answer to that is yes, does the giving of an instruction make time at large, because the architect has no power to give an extension of time?
- If the answer to that is no, is the contractor entitled to a net or gross extension of time? In other words, is the contractor entitled to an extension of time from the date for completion to the date it would actually take him to complete the variation (gross) or an extension which simply reflects the time to carry out the variation added onto the completion date (net).

In the last edition of this book, John Parris examined the first question in detail and came to the conclusion that the architect was entitled to instruct variations after the date for completion had passed. The reasoning is worth repeating.

Clause 13.2 gives the architect an unqualified power to order variations at any time. Clause 25.4.5.1 gives him power to extend the date for completion for the contractor's compliance with clause 13.2 as a relevant event. Clause 25.3.3 requires the architect not later than 12 weeks from the date of practical completion to fix a completion date *later* than that previously fixed, if fair and reasonable having regard to any of the relevant events. This can be done 'whether upon reviewing a previous decision ... and whether or not the Relevant Event has been specifically notified by the Contractor'. From that it is clear that the architect is entitled to extend the date for completion for variations ordered after the date for completion has passed and even if he has issued a non-completion certificate under clause 24.1. Clause 24.1 provides for him to issue a new certificate after a further extension of time, thus preserving the employer's right to liquidated damages.

Clearly, it follows that time is not made at large. There remains the question of the correct principle to apply to extensions of time. This point was considered in *Balfour Beatty Ltd* v. *Chestermount Properties Ltd* (1993), where it was held, under an amended version of JCT 80, that the contractor was entitled to a net extension of time as a result. The decision has not been without criticism, but it has not yet been challenged in another case in the Court of Appeal. That may be because, although the contractual reasoning may be open to criticism, the result has a certain reasonable attraction.

CHAPTER EIGHT
INSURANCE PROVISIONS

8.1   General

The part of the contract usually referred to as 'insurance provisions' is actually somewhat wider in scope than that name suggests. However, it is proposed to continue to use the name because it is a convenient way to group a number of insurance and related matters. The relevant clauses considered in this chapter are clauses 20, 21, 22, 22A, 22B, 22C, 22D and 22FC. The contents may be broadly split as follows:

- Indemnity in respect of injury to persons and property
- Insurance in respect of injury to persons and property
- Insurance in respect of the employer's liability
- Insurance of the Works: new Works and existing structures
- Insurance against loss of liquidated damages
- The Joint Fire Code

8.2   Contractor's indemnity to employer for personal injuries and death

Clause 20.1 provides an indemnity by the contractor to the employer in respect of claims arising from personal injuries to anybody or the death of anybody 'arising out of or in the course of or caused by the carrying out of the Works' unless and to the extent that it is due to the act or neglect of the employer or anyone for whom he is responsible.

In spite of the fact that the contractor is normally in possession of the site, such possession may not be exclusive of the employer's possession and the employer may, therefore, be sued under the Occupiers Liability Act 1957 or at common law or vicariously, as being responsible for the acts or omissions of the contractor, even though he is an independent contractor and not a servant. The contractor is also liable if the injury to persons is a result of breach of statutory duty.

Thus, if a person is injured due to the works, the contractor is liable except to the extent that it is the employer's fault. It is clear that the contractor is still obliged to indemnify the employer against claims for personal injury or death even if the employer's neglect is partially responsible. In such a case, of course, the contractor's liability would be reduced accordingly. An indemnity is, of course, only operative after the employer has been condemned by a judgment against him. But the employer can take out third party proceedings against the contractor before judgment is actually given against him: *County & District Properties* v. *Jenner* (1974).

From this indemnity, the clause exempts injuries or death to the extent due to any act or neglect of the employer. That is fair enough, and it is also fair enough that it should extend to acts of negligence by the employer's own servants for whom he is vicariously liable. But it also exonerates the contractor in respect of persons who are not the employer's servants, but those 'for whom the Employer is responsible'.

These may fall into two categories: those for whom the employer is deemed to be responsible under the express terms of the contract and others for whom the employer in fact and in law is vicariously responsible. Those falling within the first category are summed up by clause 29.3 which states that 'every person employed or otherwise engaged' by the employer is deemed to be a person for whom the employer is responsible for clause 20 purposes. If the judgment in *Henry Boot* v. *Central Lancashire New Town Development Corporation* (1980) is correct, that description sometimes can include statutory undertakers doing work outside their statutory obligations. Therefore, are the architect, the quantity surveyor and the clerk of works persons for whom the employer is responsible? It appears that the answer must be yes.

In practice, a person suffering injury to his person would usually claim against the employer who would, by virtue of this clause, join the contractor as a third party in any proceedings.

It may be thought superfluous to have an indemnity clause when the following clause requires the contractor to take out insurance against just the same liabilities. But if a claim was successful against the employer and the insurance company refused to meet the claim for some reason, the contractor would retain liability. The level of insurance required is to be inserted in the appendix.

8.3 *Contractor's indemnity for injury to property*

Under clause 20.2, the contractor indemnifies the employer and takes liability in the case of any loss, expense, claim or proceedings as a result of carrying out the works in respect of injury or damage to property of all kinds except the Works and site materials to the extent that it is due to the negligence or default of the contractor, his servants or agents.

The contractor is also liable if the injury to property is a result of a breach of statutory duty and, in addition to his servants or agents, he is made liable for 'any person employed or engaged upon or in connection with the Works' and any other person who may properly be on site in connection with the works excluding those persons for whom the employer is responsible.

Therefore, if property is damaged, the contractor is liable only if it is his fault: *City of Manchester* v. *Fram Gerrard Ltd* (1974). The contractor must indemnify the employer against claims for injury to property even if the employer's neglect is partially responsible. In such a case, of course, the contractor's liability would be reduced proportionately.

A person suffering injury to his property would usually claim against the employer who would, by virtue of this clause, join the contractor as a third party in any proceedings. The contractor is not liable under this clause for any loss or damage to the Works unless they have been taken into partial possession under JCT 98 clause 18 or a certificate of practical completion has been issued or determination has taken place (clause 20.3).

A question may arise whether this clause covers trespass to adjacent land or air space, e.g. by an overhead crane, removal of a right of support, nuisance to adjacent land; or escape of materials, as in *Rylands* v. *Fletcher* (1866). Trespass to land is actionable *per se*, without proof of damage or fault. Trespass to goods requires fault on the part of the person committing the tort: *National Coal Board* v. *Evans* (1951). Nuisance was at one time not thought to be actionable unless knowledge or negligence was proved. It now appears to be accepted that neither is necessary before an action will lie: *Dodd* v. *Canterbury City Council* (1979); *Lord Advocate* v. *Reo Starkis Organisation Ltd* (1981), except in the case of things naturally on the land.

There can be no doubt that *Rylands* v. *Fletcher* liability is strict, regardless of negligence or fault.

The question, therefore, is whether under this clause, the contractor indemnifies the employer if the employer is held liable for any of these torts. It all depends on the word 'default' in this context. The word was defined by Mr Justice Parker in *Re Bayley-Worthington and Cohen's Contract* (1909):

> 'Default must involve either not doing what you ought or doing what you ought not to do, having regard to your relations with other parties concerned in the transaction; in other words, it involves the breach of some duty you owe to another or others. It refers to personal conduct and is not the same thing as breach of contract.'

This approach was adopted by Mr Justice Kerr in *City of Manchester* v. *Fram Gerard* (1974). He added:

> 'Default would be established if one person covered by the clause either did not do what he ought to have done or did what he ought not to have done in the circumstances, provided . . . that the conduct in question involves something in the nature of a breach of duty.'

On the facts, he held that the conduct of sub-contractors in applying and using a waterproof coating which contained a phenolic substance and misinforming the claimants about the curing period amounted to default in the context of the indemnity clause.

More recently and perhaps surprisingly, in *Perar BV* v. *General Surety and Guarantee Co Ltd* (1994), the Court of Appeal held that default in a contractual document means a breach of contract especially if damages are said to be incurred.

It is clear that while breach of contract and breach of duty may be comprehended in the word 'default', as also may be breaches of statutory duty such as those contained in the Building Regulations and the CDM Regulations, the word is probably wider than that and includes breaches of duties which are not actionable in contract or in tort.

Since the only indemnity given to the employer under clause 20.2 is that in respect of negligence, breach of statutory duty, omission or default, it would appear that the employer is without protection under this clause in respect of the very things for which he needs protection, i.e. torts such as nuisance whereby damage is caused to adjoining or adjacent property by reason of work on his own land where liability exists independent of fault.

8.4 *Insurance against injury to persons or property: clause 21.1*

The whole of clauses 20 and 21 are subject to clause 21.3 which provides that the contractor is not liable to indemnify or insure in respect of an excepted risk as defined in clause 1.3.

Clause 21.1 requires the contractor to take out insurance for the claims noted in clauses 20.1 and 20.2. This is said to be without prejudice to the contractor's obligation to indemnify the employer as set out in clause 20. What this means in plain terms is that the contractor will be liable to indemnify the employer for such claims whether or not he takes out insurance. The taking out of insurance simply assists by ensuring that there is money available to back up that indemnity.

In addition, clause 21.1.1.2 makes clear that the insurance for claims for personal injury or death, arising in the course of that person's employment, of anyone under a contract of service or apprenticeship, i.e. any employee, with the contractor should comply with the relevant legislation.

The contractor must send the architect, on the employer's reasonable request, details of the policies and premium receipts. The architect must pass them to the employer, but it seems that he is not simply entitled to act as a 'post-box'. He has three options:

- To give advice to the employer about the documents (the architect is not usually equipped to do that); *or*
- To obtain independent specialist insurance advice and pass it to the employer; *or*
- To advise the employer to seek his own specialist insurance advice.

The last is probably the best option: *Pozzolanic Lytag* v. *Brian Hobson Associates* (1999).

If the contractor fails to insure, the employer may do so, deducting the relevant premiums from any money due to the contractor.

8.5 *Liability under indemnities in general*

An intriguing thought is whether the indemnity is effective to cover injuries which result in part from the conduct of the employer or his servants or agents.

There is authority that these words do not exempt the contractor for any 'act or neglect' on the part of the employer: *Hosking* v. *De Haviland Ltd* (1949). However, a claim against the employer for breach of statutory duty has to be indemnified by the contractor: *Murfin* v. *United Steel* (1957).

The indemnity in this clause may be valueless to the employer, because it is an established principle of indemnity law that unless an indemnity clause expressly covers the negligence of the party being indemnified, the presumption is that it is not intended to saddle the one giving the indemnity with responsibility for the indemnified's own negligence which in part contributed to the loss: *Alderslade* v. *Hendon Laundry* (1945); *Smith* v. *South Wales Switchgear* (1978). This principle has been applied to construction industry cases in *Walters* v. *Whessoe Ltd and Shell* (1960), to which reference was made in *AMF (International)* v. *Magnet Bowling and Another* (1968) which dealt with clause 14(b) of the 1957 RIBA contract, which was on similar terms to JCT 98 clause 20. There it was held that, because the contractor and the building owner were found liable as joint tortfeasors under the Occupiers Liability Act 1957 to the plaintiff, assessed under the Law Reform Act 1935 as 40% employer and 60% contractor, the indemnity clause was of no effect. It has been suggested that this decision is not consistent with the House of Lords' decision in *White* v. *Tarmac Civil Engineering* (1967) in which plant hirers were held entirely liable under the contract, even though the injury was in part the fault of the plant owner. However, the contrary is in fact the case since counsel did not rely upon the indemnity clause in view of the *Walters* v. *Whessoe Ltd and Shell* case which was regarded as authoritative.

The position appears to be that all indemnity clauses are to be construed strictly as exemption clauses *contra proferentem* the party in whose favour the indemnity is granted: *AMF* v. *Magnet Bowling Ltd and Trentham Ltd* (1968); and *City of Manchester* v. *Fram Gerard Ltd* (1974).

If the indemnity clause purports to hold a party liable for defaults, etc. other than his own, the person indemnified cannot rely on it unless it spells this out so as to make it quite clear that the one giving the indemnity is to be responsible for those over whom he has no control: *Gillespie Brothers* v. *Roy Bowles Transport* (1973).

The position may be further complicated by the Civil Liability (Contribution) Act 1978 which provides in section 7(3):

'The right to recover contributions in accordance with section 1 above supersedes any right, other than an express contractual right, to recover contributions (as distinct from indemnity) otherwise than under this Act in corresponding circumstances; but nothing in this Act shall effect

(a) any express or implied contractual or other right to indemnity; or

(b) any express contractual provision regulating or excluding contributions;

which would be enforceable apart from this Act (or render enforceable any agreement for indemnity which would not be enforceable apart from this Act.)'

If it is thought that the indemnity in clause 20 may be unenforceable for the reasons given above, this section does not in any way alter the court's power to assess such contribution between the parties 'as may be found by the court to be just and equitable having regard to that person's responsibility for the damage in question, including exempting them from all liability or directing a complete indemnity': section 2, Civil Liability (Contribution) Act 1978.

In these circumstances, it is difficult to see what value in any circumstances clause 20 is to the employer, except perhaps to extend the contractor's possible period of contractual liability under the Limitation Act 1980.

A similar indemnity to that contained in clause 20 is given by each nominated sub-contractor to the contractor in respect of sub-contract works in NSC/C clause 6.3; the same considerations apply to it.

8.6   *Insurance requirements for clause 21.2.1*

This clause originated as the result of the case of *Gold* v. *Patman and Fotheringham* (1958), which was concerned with what was to become the very well known clause 19(2)(a) of the 1939 RIBA form of contract (1952 revision). Piling sub-contractors, without any negligence, withdrew support from adjacent land and the employer was held liable in an action for nuisance. It had been argued that, by reason of the indemnity clauses, there was an implied obligation that the insurance of adjacent premises should cover both the employer and the contractor; this argument was rejected by the court.

Clause 21.2.1 clearly now puts the onus on the employer to decide whether this type of insurance is needed. If he thinks it may be, a note to that effect must be inserted in the appendix together with the amount of insurance cover which may be needed. When the contract is let, the architect, if he thinks it appropriate, may instruct the contractor to take out this insurance. It must be in the names of both employer and contractor and must cover against expense, liability, loss, claim or proceedings due to injury or damage to any property caused by collapse, subsidence, heave, vibration, weakening or removal of support or the lowering of ground water. There are nine exceptions to the cover, injury or damage:

- For which the contractor is already liable under clause 20.2
- Caused by errors and omissions in designing the Works
- Which is inevitable, having regard to the kind of work and the method of carrying it out
- For which the employer must insure under clause 22C.1 **[8.10]**.
- To the Works and site materials until practical completion
- Arising from war damage and damage caused by civil war and the like
- Directly or indirectly resulting from the excepted risks. Such risks are defined in clause 1.3
- Directly or indirectly due to pollution or contamination occurring during the insurance period unless caused by an unexpected incident occurring at a precise moment
- Due to the employer's breach of contract.

There is great scope for argument about injury or damage 'which can reasonably be foreseen to be inevitable having regard to the nature of the work to be executed and the manner of its execution' (clause 21.2.1.3). In one sense, it can be said that all excavations and pile driving inevitably create the risk of some damage, however trifling, to adjacent or adjoining properties; even the removal of an existing building can lead to an alteration in water levels in the ground. The exclusion appears to be directed at the kind of damage that everyone is aware will happen even though actual negligence is absent. For example, a house built on a raft on compacted sand will almost inevitably suffer some degree of subsidence if the retaining wall supporting the sand is removed.

The insurers must be approved by the employer, who keeps the policies and premium receipts received via the architect. Clause 21.2.3 provides for the cost of insurance to be added to the contract sum.

8.7   The alternative insurance provisions

Insurance risks in the contract are of two types:

- 'Specified perils' insurance – insurance previously known as 'clause 22 perils'. The perils are fully defined in clause 1.3 and are referred to when the insurance is to provide for fire, lightning, explosion, etc.
- 'All risks insurance' – insurance, defined at some length in clause 22.2, against physical loss or damage to work executed and site materials and against the reasonable cost of removal of debris, shoring, etc. of the Works resulting from physical loss or damage, excluding the cost of repairing, replacing or rectifying property which is defective due to wear and tear, obsolescence, deterioration, rust or mildew, any work executed or materials lost or damaged as a result of its own defect in design, plan, specification, materials or workmanship or any other work executed which is lost or damaged if such work relied for its support or stability on the defective work. Other exclusions include loss or damage arising from war, nationalisation or order of any government or local authority, disappearance or shortage only revealed on the making of an inventory and not traceable to any identifiable event, the excepted risks defined in clause 1.3. Therefore, risks such as impact, subsidence, theft and vandalism are included in this type of insurance.

   The definition had certain other exclusions applicable only to Northern Ireland (civil commotion, unlawful and malicious acts, terrorism, etc.). They were removed by JCT Amendment 3, but they are replaced by the RSUA Adaptation Schedule applicable in Northern Ireland.

The amendment also revised the definition of 'terrorism' which hitherto had meant the use of violence for political ends and included the use of violence for the purpose of putting any section of the public in fear. That is still the meaning applicable in Northern Ireland. The new definition for use elsewhere refers to the act of anyone acting in connection with an organisation directed to influencing or overthrowing a government by force.

The remainder of the lengthy clause 22 deals with insurance of the Works and existing structures. Clause 22A deals with the insurance of new Works if the contractor is to insure, clause 22B deals with insurance of new Works if the employer is to insure and clause 22C deals with insurance of the Works and of existing structures and contents. The insurances involving existing structures are always taken out by the employer. There are several alternative clauses not included in the standard form, but which have been drafted for circumstances where the parties require a different insurance scenario. There is no space (and little inclination) to discuss these somewhat rare clauses here. Suffice to say that when venturing away from the 'normal' insurance clauses in favour of these alternatives or indeed a specially drafted clause, great care must be taken that the parties are not left inadequately or, more likely, inappropriately insured.

A very important protection for sub-contractors is contained in clause 22.3.1. The contractor or employer, as appropriate, must ensure that the joint names policy either:

- Provides for recognition of each nominated sub-contractor as an insured; *or*
- Includes a waiver by the insurers of the right of subrogation against any nominated sub-contractor in respect of specified perils damage, to the Works or site materials where clauses 22A, 22B or 22C.2 apply, to existing structures where clause 22C.1 applies.

The latter cover is generally less than afforded to the employer and the contractor. Domestic sub-contractors are similarly protected except for damage occurring to existing buildings under clause 22C.1.

8.8   New Works if the contractor insures: clause 22A

Clause 22A.1 obliges the contractor to take out and maintain a joint names policy for all risks insurance. Footnote [ff] advises that some of the risks may not be possible to cover.

The value must cover full reinstatement of the works and the amount of any professional fees inserted in the appendix. Professional fees are the fees required by the construction professionals involved in the reconstruction work. The employer is entitled to deduct from insurance proceeds the amount incurred for professional fees. A strict reading of the wording of clause 22A.1 suggests that if the percentage is omitted from the appendix, the employer would be obliged to pay the fees himself. The reinstatement value should be carefully considered. If the building is effectively a total loss at a point when 50% of the work has been completed, the cost of demolition of what remains together with reconstruction at inflated prices could result in the contractor having to subsidise the project. The contractor should get very good advice from his broker before

taking out insurance to cover this risk. It does not include consequential loss: *Kruger Tissue (Industrial) Ltd* v. *Frank Galliers Ltd and
DMC Industrial Roofing & Cladding Services and H. & H. Construction*
(1998). A very important feature of the policy, indeed of all the
'Works' policies, is that they are to be in joint names. The provision
is widely misunderstood and sometimes ignored. The point is that
both employer and contractor are stated as the insured. Therefore,
once an insurance company has paid out on a claim from one of
them, it cannot exercise its usual right of subrogation (stepping into
the shoes of the insured) in order to recover against the other party.
The insurance must be maintained until practical completion of the
Works or determination of the contractor's employment even
though such determination may be the subject of dispute between
the parties.

The employer under clause 22A.2 has the right to approve the
insurers and he is entitled to have the policy documents and premium receipts. If the contractor defaults, the employer has the right
to take out the policy himself and deduct the cost from any sums
payable to the contractor.

Usually, the contractor will maintain an annual policy which
provides cover against all the risks which he may face. So the one
policy, possibly by endorsements, will include cover against liability for injury or death to persons, injury or damage to property
other than the Works, employer's liability and Works insurance.
Clause 22A.3.1 makes provision for this situation and allows the
contractor to discharge his obligations under clause 22A provided
that the policy is in joint names (a separate endorsement is
required for each contract undertaken) and that it provides cover
for not less than full reinstatement and professional fees. The contractor must provide evidence that the insurance is being maintained if the employer so requests, but there is no obligation to
deposit the policy. However, provided the request is not made
unreasonably or vexatiously, the employer may at any time
require the contractor to let the architect have the policy and premium receipts for inspection. That is not something which the
contractor would wish to do, because his annual policy is a very
valuable piece of paper. The answer is for the contractor to insist
that a suitable alternative clause is inserted before the contract is
executed, perhaps providing for a certified copy of the policy to
be handed over. The annual renewal date is to be inserted in the
appendix.

Clause 22A.4 deals with the procedure. If any loss or damage
occurs, the contractor must give written notice to the employer as

soon as he discovers the loss. A very important provision (clause 22A.4.2) makes clear that the fact that part of the Works has been damaged must be ignored when the amount payable to the contractor is being calculated. He must be paid for the work carried out although it may since have been destroyed. Therefore, the next interim payment will be certified as though the damage had not occurred.

Clause 22A.4.3 places a duty on the contractor to carry out restoration and remedial work and proceed with the Works after the insurers have carried out any inspection they require. This can cause difficulties. The contractor is not entitled to wait until he knows whether or not the claim will be accepted before he proceeds with the Works. It may take the insurers a considerable time to accept the claim. The result is often a heavy financial burden on the contractor. If the damage is very serious, the insurers may employ their own engineers and surveyors to assess the feasibility of repair or total reconstruction. The contract is completely silent about such matters, but it would be an extremely foolhardy contractor who proceeded with his own ideas of reconstruction in the face of the insurers' own views. It should also be noted that a contractor is not entitled to any extension of time if the cause of the damage lies outside those few items listed under specified perils. For example, if the building shell was erected and subsequently collapsed, the contractor would receive no extension of time for the resulting delay no matter who was ultimately at fault. However, in such circumstances, an extension of time may be the least of his worries. When serious damage occurs, it is in the interests of both parties to obtain first class advice.

Clause 22A.4.4 requires the contractor and all his sub-contractors who are recognised as insured to authorise the insurers to pay insurance monies to the employer. The contractor is entitled to be paid all the money except for any percentage noted in the appendix for professional fees. It is thought that the effect of the wording is that the employer may retain only the amount he has paid out or is legally obliged to pay out in professional fees directly related to the loss or damage, but that there is a ceiling on the amount he may retain. That ceiling is set by the percentage.

The insurance money is received by the contractor from the employer in instalments in accordance with clause 30.1.1.1. The contractor is not entitled to receive more than the insurance money and, if there has been an element of underinsurance or the policy carries an excess or if the insurers repudiate their liability, it is for the contractor to make up the shortfall (clause 22A.4.5).

8.9   *New Works if the employer insures: clause 22B*

Clause 22B provides for new building insurance to be taken out by the employer. This is not common in practice. The obligation is principally contained in clause 22B.1 and it is similar to the contractor's duties under clause 22A.1. The employer must take out and maintain a joint names policy for all risks to cover the full reinstatement value of the Works together with the percentage to cover professional fees. The employer must maintain the policy until practical completion or determination whichever is earlier. The employer must produce evidence for the contractor that the policy has been taken out and, on default, the contractor may himself take out a similar policy and he may recover the cost as an addition to the contract sum (22B.2).

The machinery for dealing with an insurance loss is in clause 22B.3. It closely follows clause 22A.4 and provides for the contractor to give written notice to the employer upon discovering loss or damage. The contractor must proceed with repairs and the execution of the Works after any inspection required by the insurers, and the contractor and his sub-contractors who are noted as insured must authorise payment of insurance monies directly to the employer. Here, however, the similarity ends. Where the employer has insured, clause 22B.3.5 stipulates that restoration, replacement and repair must be treated as if they were variations. There are two important points to note from that. First, the change does not depend on an architect's instruction. The fact that there has been loss or damage and the employer has the obligation to insure is sufficient. Second, it follows that if the repairs, etc. are to be treated as variations, they are to be valued and the employer must pay for them. This duty is not affected by any shortfall or excess in the employer's insurance nor is it affected if the insurers decide to repudiate liability. Under clause 22B, it is the employer who must make good any shortfall.

8.10   *Works in or extensions to existing structures: clause 22C*

If work is to be undertaken in an existing building or in extensions to an existing building, the appropriate clause is 22C. The insurance is to be taken out and maintained by the employer and his obligations are set out in clauses 22C.1 (existing buildings) and 22C.2 (Works in or extensions to existing buildings). Clause 22C.1 refers to a policy in joint names to cover the existing building and contents for specified perils only. The employer may well wish to have more extensive insurance and there is nothing in the contract to prevent him so doing. The contents are those which are owned by the employer or for which he is responsible. This is presumably intended to cover his goods, goods on the premises with his permission, but not goods which may be on the premises without his permission. Where portions of the new Works are taken into possession by the employer under clause 18, they are to be considered part of the existing building from the relevant date. This is a point which the employer must watch when taking possession of portions of the Works under clause 18. Clause 22C.2 obliges the employer to take out a joint names policy for the new Works in respect of all risks. Both sets of insurance must be taken out for full reinstatement value, but only in the case of the new Works must the professional fees percentage be added. Both insurances must be maintained until practical completion or determination.

Clause 22C.3 gives the contractor broad powers if the employer defaults. It is not used if the employer is a local authority. The contractor has the usual power to require proof that the insurances are taken out and are being maintained. In addition, in the case of default in respect of clause 22C.1 insurance, he has right of entry into the existing premises to inspect, carry out a survey and make an inventory of the existing structures and the contents. The only qualification on the contractor's power is that the right of entry and inspection is such as may be required to make the survey and inventory. This provision merits careful consideration by the employer, because a failure to insure by the employer may give rise to distinctly unwelcome, but lawful, entry by the contractor into the employer's property.

Clause 22C.4 contains the procedure if loss or damage occurs. (There is no express machinery for dealing with damage to the existing building and contents. That is regrettable, but the parties are left to their own devices. It would be useful if an appropriate procedure was devised and inserted in the contract.)

The contractor must give written notice to the employer on discovery of loss or damage. The contractor and his sub-contractors noted as insured must authorise the insurers to pay any insurance money directly to the employer. There is provision for either party to determine the contractor's employment within 28 days of the occurrence if it is just and equitable to do so [9.14]. If there is no determination or an arbitrator decides that the notice of determination should not be upheld, the procedure is much the same as clause 22B.3. The contractor is obliged to proceed after any inspection required by the insurers, but the work is to be treated as a variation for which the employer must pay. Shortfalls in insurance under this clause are again the responsibility of the employer. Under none of the three Works insurance clause options is the contractor penalised in respect of work already carried out and damaged by the insurance risk. It has been held that fire caused by the contractor's negligence is not covered by this insurance, but must be covered by the contractor's own insurance: *London Borough of Barking and Dagenham* v. *Stamford Asphalt Company* (1997). In that case, Lord Justice Auld in the Court of Appeal said:

> 'In my judgment, the two provisions are concerned with entirely different types of damage, in addition to the distinctions to which I have already referred. Condition 6.2 [the indemnity provision] governs the liability for damage culpably caused by the con- tractor. Condition 6.3B [employer's insurance of existing building and the Works] and its alternative 6.3A require insurance for certain damage not culpably caused by it.'

The case dealt with the JCT Minor Works Contract (MW 80), but there seems no reason why the principle should not be applied to JCT 98. *Scottish & Newcastle plc* v. *G.D. Construction (St Albans) Ltd* (2001), a case on IFC 84, has held that if the breaches of contract and negligence on the part of the contractor were assumed, the con- tractor was liable to the employer for damage to the existing structure of the building (a public house) and for business inter- ruption. Judge Richard Seymour was trying a preliminary issue. Damage had been caused to the existing structure of the public house and the Works by fire. The employer had an obligation to insure the existing structure under clause 6.3C.1 and the Works under clause 6.3C.2. The judge had to consider the relationship of clause 6.1.2 and clause 6.3C – the equivalent to clause 20.2 and 22C of JCT 98 respectively.

There has been a long stream of cases on the topic and the judge reviewed the authorities. The wording of these clauses has changed over time. Part of the judgment of Lord Justice Slade in *Canada Steamship Lines* v. *R* (1952) is worth repeating:

> 'The heads of loss and damage to which the clause relates are by no means restricted to loss or damage by fire. They also include loss or damage by "lightning, explosion, aircraft and other aerial devices or articles dropped therefrom". Even if damage by explosion could be caused by the Contractor's negligence,

damage by lightning, aircraft and other aerial devices could not. Accordingly, on analysis, to say that the Council must bear the risk of damage to the existing structure and contents falling within the last three mentioned categories is to do no more than state the obvious. Yet the draftsman has chosen so to state and, I think, for sufficient reason namely for the avoidance of doubt, particularly bearing in mind his reference to "the Works" to which I refer below. Now fire, no less than the impact of lightning, can occur without the negligence or fault of any human agency. If the draftsman chose to refer to a number of possible causes of damage which involve no fault on the part of anyone, I do not see why, in referring to fire, he should not be taken to have similarly had in mind damage by fire occurring without negligence on the part of the Contractor.'

8.11   *Non-availability of terrorism cover*

Amendment 3 inserted additional clauses at the end of clauses 22A, 22B and 22C to cover this eventuality which had previously been dealt with by an entirely separate amendment. The relevant clauses are 22A.5, 22B.4 and 22C.5. They are virtually identical.

If the insurers notify either the employer of the contractor that terrorism cover will not be available from a particular date (the 'effective date'), that party must immediately inform the other. The employer must then write to the contractor stating one of two things:

- That clause 22A.5.3, clause 22B.4.3 or clause 22C.5.3 as appropriate will apply from the effective date so far as physical loss or damage to work and materials due to fire or explosion caused by terrorism is concerned
- That the contractor's employment is determined on a date specified before the effective date.

Clauses 22A.5.3, 22B.4.3 and 22C.5.3 provide that if the relevant losses are sustained, the contractor must reinstate the work and materials and remove and dispose of debris which will be treated as a variation required by an architect's instruction. An important proviso prevents the employer reducing the amount payable to the contractor as a result of any act or negligence of the contractor alleged to have contributed to the physical loss or damage. The contractor must proceed with the Works.

If the contractor's employment is determined, no more retention is to be released and the provisions of clauses 28A.3, 28A.4 and 28A.5 (except 28A.5.5) will apply. In addition, under 22A and 22B but, strangely, not 22C, clause 28A.7 where relevant will apply.

Under clause 22A only, there are two additional clauses, 22A.5.4.1 and 22A.5.4.2 which provide that if the rate for terrorism is varied, the contract sum will be adjusted accordingly. Local authorities only may opt not to renew the terrorism cover, but to apply clause 22A.5.3 instead.

A further clause (22C.1A) is inserted after clause 22C.1, to deal with the situation where there is insurance for existing premises and contents. It is similar to the clauses mentioned above, but provides that the employer need not reinstate the existing premises.

8.12 Employer's loss of liquidated damages

The architect must make an extension of time if the contractor is delayed to the extent that the date for completion is exceeded by loss or damage caused by one or more of the specified perils (clause 25.4.3). The employer will then receive his building after the date for completion, but he will not receive liquidated damages for the delay. Clause 22D is intended to provide the employer with some relief in these circumstances. More cynically, it is simply more insurance which the employer can take out on payment of the appropriate premium. The insurance pays out if extensions are given on grounds of delay due to specified perils. If this clause is to apply, it must be so stated in the appendix. As soon as he reasonably can, after the contract has been entered into, the architect must either notify the contractor that no insurance will be required or he must instruct the contractor to obtain a quotation and for that purpose provide from the employer whatever information the contractor reasonably requires. The contractor must act as soon as reasonably practicable to send the quotation to the employer, who must instruct acceptance or otherwise. If the employer wishes to accept, the architect must so instruct and the contractor must deposit the insurance policy, together with premium receipts, with the architect for transference to the employer.

The insurance is to be on an agreed value basis, the value to be the amount of liquidated damages at the rate stated in appendix 1. The purpose of agreed value is to avoid arguments with the insurers when called upon to pay out, because they would normally expect to pay only the amount of actual loss. A footnote to the clause points out that insurers will normally reserve the right to be satisfied that the amount of liquidated damages does not exceed a genuine pre-estimate of the damages which the employer considers he will suffer. The period of liquidated damages must also be stated in the appendix. Thus, if the stated period is five weeks and the liquidated damages are set at £50 per week, an extension of time of two weeks for relevant event 25.4.3 will enable the employer to claim £100 from the insurers. If the extension is for six weeks, only £250 can be claimed, because five weeks is the limit of insurance (clause 22D.2). Clause 22D.4 permits the employer to take out the insurance himself if the contractor defaults.

8.13 The Joint Fire Code

The Joint Fire Code is dealt with under clause 22FC. Clause 1.3 defines the Joint Fire Code as:

'the Joint Code of Practice on the Protection from Fire of Construction Sites and Buildings Undergoing Renovation which is published by the Construction Confederation, and the Fire Protection Association with the support of the Association of British Insurers, the Chief and Assistant Chief Fire Officers Association and the London Fire Brigade which is current at the Base Date, and as may be amended/revised from time to time.'

The code makes clear that non-compliance could result in insurance ceasing to be available. If the code is to apply, the appendix should be completed appropriately. If the insurer categorises the Works as a 'Large Project', special considerations apply and the appendix must record that also.

Clause 22FC.2 requires both employer and contractor and anyone employed by them and anyone on the Works including local authorities or statutory undertakers to comply with the code. Clause 22FC.3 makes clear that if there is a breach of the code, the insurer may give notice requiring remedial measures, to either the employer or the contractor.

There are two situations. If the remedial measures concern the contractor's obligation to carry out and complete the Works, the contractor must ensure the measures are carried out by the date specified in the notice. However, if the remedial works necessitate a variation, the architect must issue instructions. There is provision for the contractor to act in an emergency. If architect's instructions requiring a variation are not involved and the contractor does not begin the remedial measures in seven days from receipt of the notice, or if he fails to proceed regularly and diligently, the employer may employ and pay others to do the work. In principle, the employer is entitled to recover the cost by deduction or as a debt in the usual way. Where an architect's instruction is involved, a seven day compliance notice under clause 4.1.2 may be sent in the normal way. Clause 22FC.5 provides that if the code is amended after the base date and the contractor is put to additional cost in complying, whether such cost must be added to the contract sum will depend upon whether the employer or the contractor is to take the risk. This information must be stated in the appendix.

CHAPTER NINE
DETERMINATION BEFORE COMPLETION

9.1   Discharge before completion

Under the general law, a contract can be brought to an end in several ways:

- By performance
- By agreement
- By frustration
- By breach and its acceptance
- By operation of law
- By novation.

9.1.1   Performance

This is the best way of bringing a contract to an end, because both parties have carried out their obligations under the contract and nothing further remains to be done. The purpose for which the contract was created is satisfied and the contractual relationship ceases.

9.1.2   Agreement

The parties may agree to bring the contract to an end. The only safe way to do that is to enter into another contract whose sole purpose is to end the first contract. In most cases, when a contract is ended by mutual agreement it is because each party gains something from so doing, thus satisfying the requirement for consideration as an essential element of the contract. However, sometimes proper consideration may be absent and it is wise for the parties to execute the second contract as a deed, thus avoiding any question of consideration arising.

9.1.3 Frustration

The definition of frustration was given by Lord Radcliffe in *Davis Contractors Ltd* v. *Fareham Urban District Council* (1956):

'[It] occurs wherever the law recognises that without default of either party a contractual obligation has become incapable of being performed because the circumstances in which performance is called for would render it a thing radically different from that which was undertaken by the contract.'

The fact that a contractor experiences greater difficulty in carrying out the contract or that it costs him far more than he could reasonably have expected is not sufficient ground for frustration. Neither will a contract be frustrated by the occurrence of some event which the contract itself contemplated and for which it made provision: *Wates* v. *Greater London Council* (1983). The total destruction of premises by fire was held in *Appleby* v. *Myers* (1867) to frustrate an installation contract. Extreme delay through circumstances outside the control of the parties may frustrate a building contract, but only if the delay is of a character entirely different from anything contemplated by the contract.

Where a contract is discharged by frustration, both parties are excused from further performance and the position is governed by the Law Reform (Frustrated Contracts) Act 1943. Money paid under the contract is recoverable, but if the party to whom sums were paid or payable has incurred expenses, or has acquired a valuable benefit, the court has a discretion as to what should be paid or be recoverable. In practice, it is very rare for a contract to be frustrated.

9.1.4 Breach

A breach of contract which is capable of bringing the contract to an end must strike at the very root of the contract: *Photo Production* v. *Securicor* (1980). The offending party must clearly demonstrate that he does not intend to accept his obligations under the contract. Such an instance under a building contract could take place where the client prevents the architect from issuing extensions of time and financial certificates and engages another architect to complete the work. That would be a very clear repudiation by the client. Acts of repudiation are often less clear.

Not every breach of contract by one party will entitle the other to refuse to perform his own obligations; the breach must be such as makes clear an intention to repudiate the whole of the contractual obligations. Such an intention may be conveyed by express words or by 'repudiatory conduct' which effectively prevents the other party performing what he has promised to do. For example, this happens if the employer refuses or delays to hand over the site to the contractor: *Smart & Co* v. *Rhodesia Machine Tools* (1950). If the employer failed to make provision for the necessary production information to be given to the contractor, this would inevitably be held to be a breach of contract which would discharge the contractor from further performance.

'An unaccepted repudiation is writ on water' is how one judge described the situation when one party has committed a fundamental breach. The innocent party can either accept this repudiation and sue at once for damages; or he can continue to perform his obligations under the contract and hold the other party liable for all the money due under the contract: *White & Carter* v. *McGregor* (1961). This, of course, applies only if the innocent party is capable of performing his own obligations under the contract. In the case of contractors who are refused possession of the site, clearly they have no option but to accept the repudiation, because in addition to the contractual right, they also need a property right – a licence to enter upon and occupy the site.

Failure by the employer to pay sums certified by the architect is a breach of contract. However, it is rarely taken as showing an intention to repudiate. The law regarding progress payments and payment by instalments was laid down by the House of Lords in *Mersey Steel & Iron Co* v. *Naylor Benzon & Co* (1884). Failure to pay for one or more previous deliveries did not exonerate the other party from the obligation to deliver subsequent instalments of goods. Even express refusal to pay (as in this case, on legal advice) was held not to demonstrate an intention to repudiate the contract. 'The law of England is that ... payment for previous instalments is not normally a condition precedent to the liability to deliver – although it can be made so by express provision,' said Mr Justice Cooke in the New Zealand case of *Canterbury Pipelines Ltd* v. *Christchurch Drainage Board* (1979). To apply this to the JCT 98 situation: failure or refusal to pay sums certified as due as interim payments will not in itself give the contractor the right at common law to cease to perform his obligations under the contract. His only remedies are to sue on the interim certificate or to exercise the specific provision made in JCT 98 clause 28.2.1.1. However, repeated failure to pay which causes the other party to lose all confidence that he will ever be paid may amount to a repudiatory breach: *D.R. Bradley (Cable Jointing) Ltd* v. *Jefco Mechanical Services* (1989).

9.1.5 Operation of law

This may occur if one of the parties becomes bankrupt or sufficient time passes to stop the contract remaining effective. How much time will depend on all the circumstances. A somewhat rare example of discharge by operation of law would be if the object of the contract became illegal during its currency.

9.1.6 Novation

Replacement of one contract by another usually accompanied by a change in the identity of one of the parties. In the case of a simple contract for a lump sum, if one party issues instructions to vary the contract Works, the other party is entitled to consider the original contract at an end and a new contract, incorporating the variation, in being. Severe financial repercussions may result. The effect is avoided in the standard forms of building contract by the insertion of a variation clause to allow variations of the original contract Works.

9.2 Determination without breach

The JCT 98 contract in clause 22C.4.3.1 and the excessively complicated clauses 27, 28 and 28A provides for determination of the contractor's employment (but not the contract) on the happening of events which are not repudiatory breaches of contract or, indeed in some cases, breaches at all.

None of these clauses make the events breaches of contract and in spite of the words 'without prejudice to any other rights and remedies which the Employer may possess' (clause 27.8) or '... the Contractor may possess' (clause 28.5), neither in fact has any remedies other than those expressly conferred by the contract: *Thomas Feather & Co (Bradford) Ltd* v. *Keighley Corporation* (1953).

The existence of express provisions for determination in a contract does not exclude common law rights and there is no condition to be implied to that effect: *Architectural Installation Services Ltd* v. *James Gibbons* (1989). The court held that failure to give the necessary notices under the appropriate provisions meant that a telex determining the claimant's employment, which read:

> 'By reason of your withdrawal of labour from the above contract without sufficient notice to ourselves we are obliged to hereby give notice of the termination of your contract.'

constituted an unlawful repudiation of the contract by the defendants. But it was said to be open to the defendants to produce evidence which would justify termination of the contract at common law. The case was concerned only with two preliminary issues. A passage from *Chitty on Contracts*, 25th edition, paragraph 1503 (now paragraph 23.46 of the 28th edition) was quoted with approval:

> 'The fact that one party is contractually entitled to terminate the agreement in the event of a breach by the other party does not preclude that party from treating the agreement as discharged by reason of the other's repudiation or breach of condition, unless the agreement itself expressly or impliedly provides that it can only be terminated by exercise of the contractual right.'

In spite of this judgment, there are grave difficulties in the way of a party asserting both his rights under the contract, say under JCT 98 clause 27, 28 or 28A, and claiming that the contract has been repudiated. To determine the contractor's employment under any of those clauses is in fact to *affirm* the contract, because one is using the contractual mechanism to carry out the determination. However, in the case of a repudiation, by notice or conduct, which is alleged to discharge the other party from further performance, it is effective only if the repudiation is accepted.

This point does not appear to have been taken in any of the relevant cases, but it is thought that where a default is both an event which gives a right to determination under JCT 98 clause 27 and also a repudiation at common law, the employer must elect whether he pursues his remedy under the contract: if so, he is affirming the contract. Alternatively, he may assert that the contract is at an end because of the other party's repudiation of it and his acceptance of that repudiation.

The two positions are so inconsistent that it is difficult to see how a party can assert both at the same time, even if one is contended as an alternative to the other. Whether the employer has affirmed the contract is a matter of fact; if he relies on a term of the contract, he is plainly affirming it and he cannot be heard to say that he has accepted the contractor's repudiation.

On this point it appears that the *Architectural Installation* case is not correct: *Fercometal* v. *Mediterranean Shipping* (1989). However, it does seem possible to give notice under clause 27 and then subsequently to assert that some other conduct of the contractor amounted to a repudiatory breach which was accepted. Alternatively, it may be possible to frame the notice in such a way as to assert a repudiatory breach, which was accepted, but if it was not accepted to rely upon the contractual provisions.

9.3 Notices

The general requirements concerning the calculation of days in any notice period are given in clause 1.8 **[1.11]**. However, additional requirements in the case of determination are set out in clauses 27.1, 28.1 and 28A.1.1. The notices given in relation to determination must be given in writing by actual, special or recorded delivery. In the case of special or recorded delivery, receipt will be deemed to be received 48 hours after posting. Strangely, this is expressed to be subject to proof to the contrary. In the case of such officially recorded means of posting, such proof is always available (although not, it must be admitted, always at the time such proof is required). Despite the contents of clause 1.8, the 48 hours excludes Saturdays and Sundays as well as public holidays.

The notice provision after loss or damage in clause 22C.4.3.1 has no provision for actual delivery. There is no 48 hour deeming provision for posting.

The individual determination provisions are examined below, but they do not state precisely what any notice must contain. Although there is some authority that notice may be sufficient if given in general terms (*Supermarl Ltd* v. *Federated Homes Ltd* (1981)), those charged with giving notices under the contract would be wise to ensure that they conform precisely with the contract requirements even to the extent of repeating the exact words of the contract where appropriate.

Under clause 27.2.4, 28.2.5 and 28A.1.3, where the notice is a notice of determination, it must not be given unreasonably or vexatiously. In considering whether a contractor had acted 'unreasonably' in determining in *Hill* v. *London Borough of Camden* (1980), the judge said:

'I imagine that it is meant to protect an employee who is a day out of time in payment, or whose cheque is in the post, or perhaps because the bank has closed or there has been a delay in clearing the cheque or something – something purely accidental or purely incidental so that the court could see that the contractor was taking advantage of the other side in circumstances in which, from a business point of view, it would be totally unfair and almost smacking of sharp practice.

I can think of no other sensible construction of the word "unreasonably" in this context.'

In *John Jarvis Ltd* v. *Rockdale Housing Association Ltd* (1986), to take action vexatiously was said to mean taking action with an ulterior motive to oppress or annoy.

9.4   Employer's rights to determine – grounds: clause 27.2.1

The employer's rights to determine fall into two very distinct categories. The first is what might be termed 'defaults of the contractor'; the second refers to various forms of insolvency on the part of the contractor. In addition, there is corruption. The first category is prefaced with the words in clause 27.2.1: 'If before the date of Practical Completion...' making it clear that the defaults listed can only form grounds for determination during the period before practical completion. After practical completion the grounds disappear. The contractor's defaults which constitute the grounds are the following:

- *Wholly or substantially suspending the Works without reasonable cause* (clause 27.2.1.1)
  The operative words are 'wholly or substantially'. In order for the ground to bite, the contractor need not have wholly suspended. It is enough if he has substantially done so. What would constitute 'substantial' in this context will very much depend on the nature and scope of the work being undertaken. It is thought that in order to qualify as 'substantial' the suspension usually would have to involve work which was critical to the progress of the Works as a whole. 'Reasonable cause' is not limited to the relevant events under clause 25.4. It would certainly include suspension under clause 30.1.4, but it could also include anything which could cause the contractor to suspend. For example, a serious breakdown in the contractor's machinery could be a

reasonable cause of suspension although it is not something which would give the contractor the right to suspend. Suspending, because the architect is alleged to have under certified is not a 'reasonable cause' under this clause: *Lubenham Fidelities & Investment Co* v. *South Pembrokeshire District Council and the Wigley Fox Partnership* (1986).

- *Failing to proceed regularly and diligently with the Works* (clause 27.2.1.2)

  This is a failure to proceed in accordance with the requirement set out in clause 23.1.1. The contractor's programme is a good indication of his intentions, but simply failing to work in accordance with the programme is not sufficient to prove that he is failing to work regularly and diligently. To show failure, the contractor must be failing in his duty to work constantly, systematically and industriously. The architect must look into the labour on site compared to the labour which ought to be there, the plant and equipment in use, the work to be done, the time left to complete the Works, the rate of progress, whether the contractor is likely to finish on time and, if not, whether the delay is due to factors beyond his control (which may be wider than clause 25.4 relevant events). It used to be thought that provided the contractor had any kind of presence on site and that he was making some progress, determination under this ground was impossible. However, the Court of Appeal in *West Faulkner Associates* v. *London Borough of Newham* (1995) encouraged architects to take a stronger line **[3.2]**.

- *A refusal or neglect, which causes the Works to be materially affected, to comply with the architect's written notice or instruction which requires him to remove work or materials which are not in accordance with the contract* (clause 27.2.1.3)

  A written notice or instruction must mean one which complies fully and in detail with clause 8.4: *Holland Hannen & Cubitts* v. *Welsh Health Technical Services Organisation* (1981). The Works must be 'materially affected' by the refusal or neglect. Therefore, the architect and the employer cannot exercise the determination power every time the contractor is given an instruction under clause 8.4. If the Works are materially affected, they will be affected substantially or seriously. Presumably, what is intended is a situation where defective work is in danger of being covered up by other work or where there is so much remedial work to be done that the project is virtually at a standstill. There can be nothing trivial. The wording is presumably to prevent this remedy being used instead of the other available remedies, such as the architect's duty only to include, in interim certificates, work properly executed or the provision under clause 4.1.2 to give a seven day warning notice before achieving compliance with the instruction by the use of another contractor.

- *Failure to comply with clauses 19.1.1 or 19.2.2* (clause 27.2.1.4)
  This deals with the situation if the contractor attempts to assign the contract without the employer's written consent or attempts to sub-let any part of the Works without the architect's written consent. Sub-contracting is usual and it is not the intention of the ground to allow the employer to determine simply because the contractor has forgotten to notify the architect. For example, if the contractor has sub-let without consent, but when the architect finds out he has no real objection to the contractor, he would be acting unreasonably in withholding his consent and in issuing a default notice on this ground. The better way would be for the architect to ratify the situation with a letter. It is suggested that only if the contractor fails to obtain the architect's consent, the architect has firm grounds to refuse the consent and the contractor attempts to continue with the sub-contract can determination be attempted on this ground with any real prospect of success.

- *Failing to comply with the CDM Regulations in accordance with the contract* (clause 27.2.1.5)
  Clause 6A deals with compliance with the CDM Regulations by the contractor, particularly in clauses 6A.2, 6A.3 and 6A.4. No guidance is given about the extent of failure required to justify determination. Each case would depend on its own facts, but a strict approach would be supportable wherever a real danger to health or safety was threatened.

9.5 *Procedure for employer's determination: clause 27.2*

The complex procedure requires strict adherence. It is triggered by the architect giving the contractor a notice specifying the defaults. The employer cannot send this letter and it will not be effective if he does so: *Hill* v. *London Borough of Camden* (1980). Clause 27.2.2 provides that the employer (not the architect) may serve notice on the contractor determining his employment if the contractor does not stop the default within 14 days of receipt of the default notice. It is worth noting that the employer has only 10 days from the expiry of the default notice in which to serve the termination notice.

Other than the employer proceeding to serve a determination

notice, there are two possible scenarios, both of which are dealt with under clause 27.2.3. The contractor may end his default within the 14 day period or the employer in any event may opt not to determine within the 10 days. In both instances, the position is that if the contractor repeats the default already notified, the employer may serve notice of determination without the need for another default notice. Whenever the determination notice is served, it is said to take effect on the date it is received.

Care must be taken that the repeated default is the same as the default originally notified. If it is the slightest degree different, the safest course is for the architect to serve another default notice to restart the default process. Even if the default is indisputably the same as the original default, it may not be repeated until some months after the first occurrence. Although it is not strictly necessary, it would be prudent for the architect or the employer to write to the contractor before issuing the determination notice, pointing out that the letter is not a default notice under clause 27.2.1, but is sent as a warning that the employer intends to exercise his rights under clause 27.2.3 if the contractor does not cease the repetition forthwith. This action would be sufficient to deflect any suggestion that the employer was acting unreasonably by determining after a repetition. The chances of the contractor being successful in any accusation of unreasonableness, in any event, would be hard to sustain in the face of the contract's unequivocal language.

9.6   *Contractor's insolvency: clause 27.3*

The second category deals with the employer's rights to determine as a result of the insolvency of the contractor. Clause 27.3.3 provides that the contract is determined automatically if any of the following events occur:

- A provisional liquidator or trustee in bankruptcy is appointed
- A winding up order is made
- A resolution for voluntary winding up is passed (unless for the purpose of amalgamation or reconstruction).

If both employer and contractor agree, the employment may be reinstated. So far as other reasons for insolvency are concerned, the employer may determine the contractor's employment by written notice under clause 27.3.4 at any time. That is subject to the provision in clause 27.5 for the parties to reach a different agreement. The insolvency events allowing the employer to determine at will are:

- A composition or arrangement with creditors
- A proposal for a voluntary arrangement for a composition of debts or scheme of arrangement to be approved under the Companies Act 1985 or the Insolvency Act 1986
- Appointment of an administrator or administrative receiver under the Insolvency Act 1986.

In some, but curiously not all, cases of insolvency, clause 27.3.2 requires the contractor to inform the employer immediately in writing. They are:

- A composition or arrangement with creditors
- A proposal for a voluntary arrangement for a composition of debts or scheme of arrangement to be approved under the Companies Act 1985 or the Insolvency Act 1986.

It is not clear why these two events have been singled out. A common problem is that the architect may suspect that the contractor (or the contractor may suspect that the employer) is on the brink of insolvency but, so far as the architect can ascertain, no recognisable insolvency event has taken place. In such circumstances, the employer is powerless to take any action under the contract.

Clause 27.5 allows the employer and the contractor to explore options. Crucially, clause 27.5.1 provides that the employer is not obliged to make any further payment to the contractor from the date on which the employer could have determined the contractor's employment under clause 27.3.4. Linked to this clause in effect is clause 27.5.4 which provides that the employer may take 'reasonable measures' to ensure that the Works, materials and the site are protected. The contract describes the standard of protection as 'adequate', in other words sufficient protection taking into account the locality, the materials to be protected, the incidence of vandalism and so on. The contractor must not hinder or delay the employer in taking such measures although the contract does not go so far as to say that the contractor must actively co-operate. The employer may deduct the reasonable cost from money due to the contractor provided the relevant withholding notice is given **[5.21]** or he may recover it as a debt.

This situation continues until the employer either exercises his right to determine (clause 27.5.2.2) or until he makes an agreement under clause 27.5.2.1. There are three options listed under clause 27.5.2.1:

- Continue the carrying out of the Works
- Novate the contract
- Conditionally novate the contract.

Once the agreement is concluded, the parties will be subject to its terms. What course the employer takes will depend on all the circumstances and the advice of the professional team and possibly legal advisors. Determination involves a fresh start with a new contractor and an inventory of the work done and materials together with a potentially complicated tendering process. Delay will be inevitable. For the contractor to continue requires an act of faith on the part of the employer and possibly some kind of guarantee. Novation of the contract can be a solution particularly if the novation replaces the contractor with its parent or sister company which may well continue to employ the existing personnel. Precisely what is envisaged by conditional novation is difficult to imagine.

Clause 27.5.3 deals with the period between the date the employer could have determined and actual determination or entering into a clause 27.5.2.1 agreement. To give the parties breathing space to decide the best course of action, they may enter into an interim arrangement to allow work to continue. An important proviso to protect the contractor's position states that any payment which falls due under such an arrangement will not be subject to set-off other than any deductions under clause 27.5.4. It must be remembered that such an arrangement would be in writing and, therefore, would be a 'construction contract' for the purposes of the Housing Grants, Construction and Regeneration Act 1996. Except in the unlikely event that the parties make express conforming provision, the Scheme for Construction Contracts (England and Wales) Regulations 1998 would apply.

9.7 *Determination for corruption: clause 27.4*

This clause is very straightforward. It provides for the employer to determine the contractor's employment under this contract or any other if the contractor has taken or given bribes to any person or if he commits any other offence in relation to this or any other contract with the employer under the Prevention of Corruption Acts 1889 to 1916 or if the employer is a local authority, under sub-section (2) of section 117 of the Local Government Act 1972. No notice period is required.

The prohibition applies, not only to the contractor, but also to any employee or anyone acting on his behalf, albeit in a temporary capacity. The fact that the contractor may be entirely unaware of the act is irrelevant. In any event, corruption is a criminal offence and there are severe penalties. Even if the contract did not expressly provide for determination for this reason, the employer would be able to rescind the contract at common law and/or recover any secret commissions. The employer would be well advised to seek immediate legal advice if corruption is suspected.

The contract appears to assume that persons for whom the employer is responsible will be whiter than white, because there is no provision for corruption the other way round, i.e. corruption of the contractor by a servant or agent of the employer.

9.8   Employer's rights on determination: clauses 27.6 and 27.7

The procedure to be followed after determination under clause 27 is set out in two parts: clause 27.6 if the employer decides to complete the Works; clause 27.7 if he decides not to complete the Works. Clause 27.7 was added following *Tern Construction Group Ltd* v. *RBS Garages Ltd* (1992) where, under JCT 80, the employer did not complete the Works and the question arose whether substantial completion was a condition precedent to the right to payment.

The employer is entitled to employ other persons to carry on the Works to completion including dealing with defects after the defects liability period **[12.4]**, and may for that purpose use the temporary plant, buildings, tools and the site materials. He may also purchase further materials. If the plant and so on is not owned by the contractor, the employer may not use them without the owner's consent. This particularly relates to hired in plant.

Surprisingly in view of the provision in the JCT Intermediate Form (IFC 98), there is no express clause requiring the contractor to give up possession of the site. The absence of such express provision has caused difficulties in the past: *London Borough of Hounslow* v. *Twickenham Garden Developments Ltd* (1970). It is possible that the contractor has the right to obtain an injunction to continue occupation of the site where determination is disputed: *Mayfield Holdings Ltd* v. *Moana Reef Ltd* (1973). The employer must remember that if the contractor is liable to insure the Works, the liability ceases on determination. Surprisingly, there is no provision (as in clause 28.4.1) for the contractor to leave the site in a safe condition, but the contractor has a duty to those he can reasonably foresee may be injured by the consequence of his actions. He must, therefore, take reasonable precautions.

If the employer so requires, the contractor must assign to the employer within 14 days of determination the benefit of any agreement for the supply of goods, or the carrying out of work, etc. provided it is assignable. This provision does not apply if the contractor is insolvent.

Another provision which does not apply if the contractor is insolvent is the employer's right to pay directly any supplier or sub-contractor for the supply of materials or the carrying out of work if not already paid by the contractor. Needless to say, such payments may be deducted from any money becoming due to the contractor or may be recovered as a debt. Clearly, it is not permissible to make the suppliers and sub-contractors into preferential creditors in the case of insolvency, but the employer may be no less dependent on the original firms after an insolvency than after any other determination event. How the employer can persuade such firms to complete the work without the incentive of being paid what they are owed, is a difficult question to answer and may require an entirely fresh approach to the problem on the part of such firms and the employer.

Clause 27.6.3 empowers the architect to require the contractor to remove all his plant, etc. from the Works. The contractor has a reasonable time in which to comply, after which the employer may remove and sell the contractor's property still on site. Any money raised after deduction of the employer's expenses must be kept by the employer to the credit of the contractor to be taken in account at the completion of the Works. The employer is expressly absolved from any liability for loss or damage which may occur during the removal. This is a useful provision and avoids any dispute about the matter at a subsequent time. It is not unusual for site accommodation to virtually disintegrate when being dismantled after a long period on site.

Clause 27.6.4.1 is very important and much misunderstood. It provides that contract provisions which require payment or release of retention to the contractor will not apply. However, that provision is subject to the following exceptions:

- Payments made under an interim clause 27.5.3 arrangement after an insolvency event are permitted.

- Clause 27.6.4.2, which requires an account to be drawn after completion of the Works and rectification of any defects after the defects liability period, will apply.
- The contractor may enforce his right to payment of any amounts which should *properly* have been discharged and which have *unreasonably* not been discharged. This right applies to amounts owing 28 days or more from the date of determination or when, under clause 27.3.4, determination could first have been notified. This clause is very subtle and the words here italicised should be carefully considered. There is scope for considerable dispute in practice.

Within a reasonable time after the finish of the making good of defects after the end of the defects liability period, the employer may prepare a statement or the architect may issue a certificate setting out the financial state of the project in accordance with clause 27.5. The employer's total expenses including those incurred in completing the Works, any direct loss and/or damage resulting from the determination, the amount of any payments to the contractor and the total amount which would have been payable under the contract had it been correctly executed, must be set out. An appropriate calculation will reveal whether an amount is owing to the contractor or to the employer. Whichever it is, the contract provides that it is to be treated as a debt.

Under clause 27.7.1 the employer has the option to decide not to complete. The employer has six months from the date of determination to decide. If he opts not to complete, he must so notify the contractor in writing. The employer must then send a statement to the contractor in which he (or probably the quantity surveyor) will set out:

- The total value of work properly executed together with any other amounts due to the contractor
- The amount of any expenses due to the employer including loss and/or damage resulting from the determination.

If the difference between these figures and the amount already paid to the contractor results in a balance in favour of either employer or contractor, the difference is to be treated as a debt owing to the appropriate party.

Clause 27.7.2 is a default provision in case the employer has done nothing after six months. The contractor may write and ask him whether he is going to complete under clause 27.6 or, if not, requiring a statement of account. If the employer remains silent, it would be a breach of contract and the contractor could take action through one of the dispute resolution procedures.

9.9   *Determination by contractor – grounds: clauses 28.1 and 28.2*

If the contractor determines his own employment, the consequences for the employer will be extremely serious. He will be obliged to look elsewhere for a contractor to complete the Works. The completion of Works after determination is not a favourite job among contractors; there are too many unknowns. A complete bill of quantities will be required, prepared after a detailed survey of what has been done on site. Almost inevitably, the cost of completion will be underestimated in some way and the employer will be faced with mounting bills. There will be additional professional fees and the employer may seriously consider withholding fees or taking legal action against the professional team. Usually, the contractor will be entitled to recover the profit he would have made if he had completed the Works: *Wraight Ltd* v. *P.H. & T. (Holdings) Ltd* (1968). The completion of the Works will be substantially delayed and the total cost will be increased.

Some of these consequences would also apply to the situation if the employer determined under clause 27. However, in that case, the employer would at least have a claim against the contractor for any loss or damage caused (albeit that would be of little use against an insolvent contractor).

There are effectively five grounds under which the contractor can determine. They are found in clause 28.2:

- *The employer does not pay amounts properly due under any certificate by the final date for payment, including VAT*
  It is noteworthy that there has to be an unpaid certificate for this ground to bite. Although the contractor may be able to make out a good case that money has not been properly certified, it will not avail him under this ground. Reference to VAT amounts are welcome. Some employers appear to look on the payment of VAT as an optional extra.
- *The employer interferes with, or obstructs, the issue of any certificate*
  It is important to remember that this ground refers to any certificate and not just a financial certificate. *R.B. Burden* v. *Swansea Corporation* was a case decided in 1957. There, Lord Somervell expressed the hope that the words used in the earlier RIBA

contract would be clarified. His wish has not been granted and the words remain the same in JCT 98. In the *Burden* case it was held that these words applied if the employer directed the architect (including his employee architect) to withhold a certificate or dictated the amount in it. Any interference with an architect's duty to act independently and fairly between the parties will be sufficient grounds and if the architect disqualifies himself by such conduct in accepting such instructions, the contractor will be able to recover without a certificate: *Hickman* v. *Roberts* (1913); *Europa* v. *Leyland* (1947). An instruction to an in-house architect to withhold a certificate until local authority auditors have approved it is clearly caught by this clause.

- *The employer fails to comply with the assignment provisions*
This is straightforward. The employer may not assign without the contractor's written consent. The only thing that the employer can assign is his rights – he can never assign his duties unless novation is employed. This clause addresses the situation in which the employer may wish to sell on the building to another. Even after an agreement assignment of the right to receive the finished building, the employer would remain liable for payment and any other duties under the contract. However, the contractor may have strong objections to any envisaged assignment of rights. The assignment would not be valid without the contractor's consent and this ground allows the contractor to determine if such assignment is attempted.

- *The employer fails to comply according to the contract with the CDM Regulations*
Clause 6A.1 deals with compliance with the CDM Regulations by the employer. His obligation is to ensure that the planning supervisor carries out his duties under the Regulations and that if the contractor is not the principal contractor that the principal contractor carries out all his duties under the Regulations. This probably amounts to a guarantee on the part of the employer.

- *The carrying out of the whole or substantially the whole of the Works is suspended for a continuous period of the length stated in the appendix due to one of four reasons*
If no period is stated in the appendix, the default period is one month. The reasons are:
  - *The contractor not having received, in accordance with the information release schedule or otherwise at the proper time, instructions and other information*
  This is a failure on the part of the architect to comply with

clause 5.4. The failure would have to be gross for a resultant suspension of the Works for one month to take place.

– *Architect's instructions issued under clause 2.3 (discrepancies), 13.2 (variations) or 23.2 (postponement) unless due to the default of the contractor*

Postponement appears to be the prime candidate for causing a prolonged suspension.

– *Failure or delay by the employer in carrying out work which does not form part of the contract or failure or delay in supplying materials which the employer has agreed to supply*

Although there is provision under clause 29 for the employer to arrange for the carrying out of work not forming part of the contract, there is no provision for him to supply materials or goods except of course in connection with clause 29 work.

– *Failure of the employer to give ingress to or egress from the site, including necessary passage over land in the possession and control of the employer after the contractor has given any notice which the contract documents require; or a failure to give ingress or egress as agreed between the architect and the contractor*

The implications of this provision are discussed in **[7.11.4]**.

9.10   *Contractor determination procedure: clauses 28.3 and 28.4*

The procedure for determination in the event of any of the five grounds being relevant is quite brief. If the employer defaults or causes the Works to be suspended as noted above, the contractor may serve a written notice specifying the default or the suspension event. If the employer does not bring the defaults or suspension events to an end within 14 days from receipt of the notice, the contractor may serve a determination notice. Like the employer, he has just ten days in which to act. If the determination is not served for any reason, the contractor may serve the notice within a reasonable time of the default being repeated. Notably, the length of any repeated suspension is not important, only that the regular progress is likely to be substantially affected.

9.11   *Employer's insolvency: clause 28.3*

Clause 28.3 deals with the employer's insolvency in shorter order than the contractor's insolvency under clause 27.3. The insolvency events listed are identical to those under clause 27.3.1.

If the employer's insolvency is due to one of the following:

- A composition or arrangement with creditors
- A proposal for a voluntary arrangement for a composition of debts or scheme of arrangement to be approved under the Companies Act 1985 or the Insolvency Act 1986.

He must immediately inform the contractor. The contractor simply serves a determination notice if any of the insolvency events have taken place. There is no requirement for any prior notice. A useful provision in clause 28.3.3 states that the contractor's obligation to carry out the Works is suspended as soon as an insolvency event occurs. Obviously, when the contractor's determination notice is received by the employer, such obligation is at an end.

9.12   Consequences of determination by the contractor: clause 28.4

The consequences of determination under clause 28 are fairly brisk. The process is without prejudice to any accrued rights or remedies of either party, or to the contractor's liability to indemnify the employer against any injury before or after the removal of temporary buildings, etc. The provisions of contract requiring payment or release of retention cease to apply.

Clause 28.4.1 states that the contractor must remove and arrange for all his sub-contractors to remove all temporary buildings, plant, site materials, etc. from site with 'all reasonable despatch', and so as to prevent injury, death or damage.

The employer has just 28 days after determination in which to release all the retention. The employer may carry out deductions from such retention before he pays, but the right to deduct must pre-date the determination. For example, he cannot make deductions in respect of money the employer expects to have to pay to finish the Works.

Clause 28.4.3 addresses the account which must be drawn up by the contractor with 'all reasonable despatch'. The contractor must include the following for himself as well as for all nominated sub-contractors:

- The total value of work properly done together with any other amounts due under the contract provisions

- Any sum ascertained for direct loss and/or expense under clauses 26 and 34.3
- Reasonable costs of removal
- The cost of materials properly ordered for the Works for which the contractor has paid or is legally bound to pay, for example, because the contractor is contractually bound to pay. Materials will have been 'properly ordered' when they have been ordered with due regard to the delivery dates and storage facilities at the date of determination.

Amounts previously paid to the contractor must be taken into account and the amount then properly due must be paid by the employer within 28 days of receipt of the account. The clause makes clear that no retention can be withheld.

9.13 Determination by either party: clause 28A

The idea of this clause appears to be to allow either the employer or the contractor to determine if the Works are suspended by events which are not the fault of either party. Therefore, the clause displays an even-handed approach. The Works must have been suspended for a continuous period shown in the appendix (default periods shown in brackets after each ground) by reason of:

- *Force majeure* (three months)
- Loss or damage by insurance risks caused by specified perils (three months)
- Civil commotion (three months)
- Architect's instructions issued under clause 2.3 (discrepancies), 13.2 (variations) or 23.2 (postponement) as a result of the negligence or default of a local authority or a statutory undertaker carrying out statutory obligations (one month)
- Hostilities involving the UK (one month)
- Terrorist activity (one month).

These grounds have been sufficiently discussed under the description of relevant events **[7.11–7.13]**.

Either party may give written notice stating that if the suspension is not brought to an end within seven days after receipt, the contractor's employment is determined.

The contractor may not give notice about loss or damage by insurance risks caused by specified perils if the cause was his own negligence or default and no notice by either party may be given unreasonably or vexatiously.

After determination clause 28A.2 brings to an end the effect of any clauses which require further payment and substitutes clauses 28A.3 to 28A.7.

Under clause 28A.3, the contractor must remove all temporary buildings and plant, etc. from the site virtually as required under clause 28.4.1 **[9.12]**. The employer must release half the retention within 28 days of determination subject to any right to deduct from it which arose before the determination (clause 28A.4). The contractor has two months in which to provide the employer with all documents for preparation of the account. No time period is set for the employer (or more likely the quantity surveyor) to prepare the account, but the words used, 'with all reasonable despatch', suggest that no time should be wasted (clause 28A.5). The amounts to be included in the account are:

- The total value of work properly executed together with any other amounts due under the contract
- Any sum ascertained, before or after determination, under clauses 26 and 34.3 as loss and/or expense
- The reasonable cost of removal
- The cost of materials properly ordered for the Works for which the contractor has paid or is legally bound to pay, for example, because the contractor is contractually bound to pay
- If the determination has occurred as a result of suspension following loss or damage to the Works due to specified perils and the cause was the employer's negligence or default: any direct loss and/or damage caused to the contractor by the determination (clause 28A.6).

Any amount already paid to the contractor is to be deducted from this total and the employer then has 28 days from its submission in which to pay the balance. In doing so, the employer must give the contractor and nominated sub-contractors written notice of the amounts due to the respective nominated contractors.

9.14   *Determination under clause 22C.4.3.1*

This clause provides for either the employer or the contractor to determine the contractor's employment within 28 days of the occurrence of loss or damage to the Works or to unfixed materials or goods caused by the risks covered in the Joint Names Policy in clauses 22C.2 and 22C.3. The clause refers to work being done by way of alteration or extension or both to existing structures, but not to loss or damage to the existing structures or contents. The contractor must give notice in writing to the architect and to the employer as soon as he discovers the damage. His notice must state the extent, nature and location of the damage. Although the 28 days begin to run from the occurrence and not from notification, in practice if the damage is likely to be such as to form the basis for determination, it will be discovered and notified immediately it occurs.

There is an important proviso, 'if it is just and equitable to do so'. This proviso goes to the heart of the matter and points to the difference between this ground for determination and the ground in clause 28A.1.1.2 which requires a three month period of suspension. What is just and equitable depends on the circumstances. It appears that the situation envisaged by the clause is one in which the degree of damage is such that it is uncertain whether work will recommence at all. A restoration contract for a large theatre which is subsequently gutted by fire might be such a situation.

Either party may refer the question to adjudication or arbitration, but the contract attempts to restrict that right in two ways: the procedure is to decide only whether the determination is just and equitable and clause 22C.4.3.1 states that it must be commenced within seven days of receiving the determination notice. The latter restriction flies in the face of the requirement in section 108(2)(a) of the Housing Grants, Construction and Regeneration Act 1996 that the contract shall 'enable a party to give notice at any time of his intention to refer a dispute to adjudication'. The Act, of course, takes precedence over the contractual provisions.

9.15 Suspension of the Works

The contract now provides for the contractor to suspend if the employer fails to pay. Without that express power, could the contractor partially suspend work, reduce his activities on site or go slow until he is paid? Certainly, some of the authorities suggest that he can.

Suspension of the Works by a contractor may in fact amount to a reaffirmation of the contract rather than an intention to repudiate it. In *F. Treliving & Co Ltd* v. *Simplex Time Recorder Co (UK) Ltd* (1981), it was argued that stopping work in the middle of a contract and refusing to go on was plainly repudiation and that for this purpose there was no difference between suspending and stopping.

The sub-contractor in that case had threatened to suspend work failing payment to him of monies he claimed were due for disruption, and the defendants had brought in another sub-contractor. The judge said:

> 'I should have thought that "suspend" eliminates the essential quality for repudiation by refusal to go on and introduces a temporary quality into the stop.
>
> I have been referred to the cases of *Sweet & Maxwell Ltd* v. *Universal News Services Ltd* (1964) and *Woodar Investment Development Ltd* v. *Wimpey Construction (UK) Co Ltd* (1980). Both these are authorities for the proposition that a party who takes action relying upon the terms of the sub-contractor and not manifesting otherwise an intention to abandon the contract is not to be treated as repudiating, even if he is wrong in his construction of the contract.
>
> That is not really this case.
>
> Looking at the conduct of Treliving objectively, I ask myself the question whether it has been shown that it was their clear intention to abandon and to refuse performance of the contract.
>
> I think the answer to that question is "no".
>
> They rightly contended that Simplex were in breach of the condition to afford them unimpeded access. Breach of a condition so fundamental to this sub-contract would have entitled Treliving to treat the sub-contract as repudiated.
>
> They did not do so, but gave Simplex the opportunity to which Simplex assented when [the Managing Director of Simplex] indicated his willingness to pay a substantial sum in part satisfaction of their claim.
>
> The failure of Simplex to do so caused, and in my view justifiably caused, Treliving to hold up performance of the work temporarily in order to give Simplex further time to meet that claim, the entitlement to which they had already acknowledged.
>
> At that time it is clear that it was the intention of Treliving to continue in due course the performance of the sub-contract.
>
> Treliving have not been shown to have repudiated the sub-contract, which was accordingly wrongfully terminated by Simplex.'

Similarly in *Hill* v. *London Borough of Camden* (1980) contractors working under a JCT 63 contract, reduced the size of the labour force on site and started removing from the site a number of pieces of equipment such as dumpers and concrete mixers apparently because they were dissatisfied over delay in the architect's certification of extra work executed under variation orders.

The defendant council, the employers, claimed that those acts evinced an intention to repudiate the contract. Lord Justice Lawton disagreed:

> 'It is impossible to say that they did anything of the kind. The one thing that they did not purport to do was to leave the site and indeed the employers have never suggested that they did.
>
> Indeed their subsequent conduct indicates ... that they were treating the contract as still subsisting. All that can be said against them is that by removing men and plant from the site in the way that they did they may not have been "regularly and diligently proceeding with the work".
>
> I see no reason why the defendants should say the plaintiffs have unlawfully repudiated the contract by what they did ... The most they can say is that they may be able to prove some small amount of damage as the result of the activities of the plaintiffs at the material date...'

Lord Justice Ormrod agreed with that:

> 'Everything the plaintiffs have done in this case has evinced not an intention not to be bound by the contract but the precise contrary and they have evinced an intention to treat the contract as still subsisting.'

Their intention to treat the contract as still subsisting was apparently evinced by the fact that they gave notice to determine under what is now clause 28.2 **[9.9]**.

Under clause 30.1.4, the contractor now has a contractual right to suspend his obligations if the employer fails to pay. The right is included in accordance with section 112 of the Housing Grants, Construction and Regeneration Act 1996, but in an important respect noted below, the clause is more restrictive than the Act. The employer's failure which triggers the contractor's right to suspend is a failure to pay in full by the final date for payment and expressly includes VAT. Therefore, the contractor has a right to suspend if the employer pays the whole of the amount certified, but only half the amount of applicable VAT. The contractor's right is to suspend subject to the employer's right to withhold payment if the correct notices are properly served **[5.21]**. If the employer has not paid in full and he has not served the correct withholding notices, the contractor may serve a written notice on the employer with a copy to the architect. The notice must state that he intends to suspend and the grounds. Although the clause does not expressly mention it, it is obviously good practice for the contractor to refer to the seven days which the employer has in which to pay. The contractor may suspend his obligations until such time as payment is made in full.

It is worth noting that suspension of obligations is rather wider than just suspension of the Works. For example, it appears that the contractor can suspend all his obligations under the contract including the obligation to keep the Works insured. Of course, the contractor would retain his ordinary obligations under the general law and under statute to ensure that the site was left in a safe condition. However, it appears that he would not remain liable for maintaining it in that condition. If a contractor does suspend, it is prudent for him to send the appropriate notices to all regulatory bodies.

Under the Act, a contractor who lawfully suspends is effectively entitled to an extension of time. However, the Act says nothing about any reimbursement for any loss or expense that the contractor may suffer as a result of the suspension and, possibly, starting up again. JCT 98 in clause 25.4.18 makes such suspension a relevant event **[7.11.8]**. It goes further. Clause 26.2.10 makes such suspension a 'matter' entitling the contractor to be reimbursed any loss and/or expense **[10.11.10]**. However, read strictly, the right to such loss and/or expense depends on the seven day suspension notice being given to the employer with a copy to the architect. If the contractor merely gives notice to 'the party in default' (i.e. the employer), he will comply with the Act, but not with the contract. In practice, it seems unlikely that the contractor will forget to copy the architect. It is much more likely that a contractor may give the notice only to the architect, forgetting the employer completely, thus neither complying with the Act nor the contract and rendering any subsequent suspension an extremely dubious action.

CHAPTER TEN
CLAIMS AND COUNTERCLAIMS

10.1 Introduction

It has been said that contractors' claims can be put into six categories:

10.1.1 Claims for loss and/or expense arising under clause 26: matters affecting the regular progress of the Works, or under clause 34.3: antiquities

The architect is authorised by the contract to deal with these claims.

10.1.2 Claims at common law for breach of the express terms of the contract by the employer or by those for whom he is vicariously liable

The architect has no authority under JCT 98 to deal with these claims and if the employer wishes to allow him to settle or compromise or otherwise deal with them, he must be given specific and separate authority – preferably in writing. The amount of any such settlement, of course, cannot be certified under the terms of the contract and must be dealt with quite separately. These claims are sometime known as 'ex-contractual' meaning 'outside the contract'. This can be confusing, because *ex-contractu* in ordinary legal terms means 'arising out of the contract'.

10.1.3 Claims at common law for breach of the implied terms of the contract

For example, in *Holland Hannen & Cubitts* v. *Welsh Health Technical Services Organisation* (1981), it was alleged that the defendants were in breach of their implied obligation, by their servants or agents, to do all things necessary to carry out and complete the Works 'expeditiously, economically and in accordance with the main contract', further that neither the employer nor the sub-contractors (who were directly nominated by the employers) would in any way 'hinder or prevent the contractor' from carrying out and completing the Works. So far as implied terms are concerned, see **[1.17]**.

10.1.4   Claims either at common law or under the statutory provision of the Misrepresentation Act 1967 for misrepresentation by the employer to the contractor

Such misrepresentation may comprise a misdescription of site conditions as in the Australian case *Morrison-Knudsen International v. Commonwealth of Australia* (1972), where the misdescription of clay subsoil failed to observe that it was full of cobbles. Similarly in *Bacal Construction (Midlands) Ltd* v. *Northampton Development Corporation* (1975) the invitation to tender described the soil as a mixture of Northamptonshire sand and upper lias clay. It contained tufa. Alternatively, the misrepresentations may consist of pre and post-contractual observations about the nature of the work, the sequence of operations and the effect of designs by the employer and his architect. *J. Jarvis and Sons Ltd* v. *Castle Wharf Developments & Others* (2001) put a slightly different slant on the position. There were complex questions concerning the extent to which planning permission had been given. The Court of Appeal found that a professional, whether architect, project manager or other, who induced a contractor to tender in reliance on misstatements from the professional, could become liable to the contractor if it could be demonstrated that the contractor relied on the misstatements. In this instance, the court held that the contractor had ample opportunity to investigate the position and did not rely on the professional.

10.1.5   Ex gratia claims

These are claims for which there is no legal justification, either under the contract or for breach of the contract. They rely on the employer parting with his money out of the kindness of his heart. There may be commercial reasons why an *ex gratia* payment may be made. At one time, highway authorities and government departments did make ex-gratia payments to some contractors for sufficient reason, perhaps where there was a fixed priced contract and rampant inflation. An employer may decide to make such a payment to prevent a contractor becoming insolvent and, therefore, saving the enormous cost and delay associated with determination of employment and the engagement of another contractor at an increased price. The making of any payment where there is the slightest hint of insolvency, however, must be subject to stringent scrutiny and appropriate safeguards.

10.1.6   Quantum meruit claims

This topic has already been described **[1.21]**. It can occur where work is done where there is no contract; where the existing contract is frustrated by supervening impossibility of performance; or where work is done outside the scope of the contract.

There are other, less flattering categorisations. They have been described as being of three sorts: justified, speculative and fraudulent. It is surprising how many contractors make claims which fall into the last category or very near it. How else can one describe a contractor's claim based on flimsy evidence and which is eventually ascertained at less than a third of its original amount?

10.2   *JCT 98 claims clauses*

As part of the overall scheme of payment, most standard form contracts make provision for the contractor to recover money which he has either lost or expended as an essential part of carrying out the contract. All such provisions place conditions on the right to recovery and they generally allow additional or alternative claims for damages for breach of contract at common law. So far as the contract machinery is concerned, the parties must take note of and observe the precise wording of the clause if the contractor is to be properly reimbursed.

Two provisions in JCT 98 may give rise to loss and/or expense claims by the contractor. They are clause 26, which deals with loss and expense caused by matters materially affecting regular progress of the works, and clause 34.3, which deals with loss and expense arising from the finding of 'fossils, antiquities and other objects of interest or value' found on or under the site.

The loss and/or expense which is the subject of the contractor's claim must be *direct* and not consequential. The listed event or events relied upon must be the cause of the loss and/or expense and the phrase may be equated with the common law right to damages.

Both of these clauses impose obligations on the contractor and on the architect and/or quantity surveyor. Each of them confers on the contractor a legally enforceable right to financial reimbursement for 'direct loss and/or expense' suffered or incurred as a result of specified matters provided that he observes the procedures laid down by the provisions. Once the claims machinery has been put in motion by the contractor, the architect and/or quantity surveyor must carry out the duties imposed on them.

However, the provisions for the giving of notice or the making of applications are not merely procedural. They are included in the contract to give the architect the opportunity to investigate the claim reasonably when the specified matters happened and to give the employer the opportunity to see if there is another way around the problem. In addition there is provision in clauses 13.4.1.2 and 13A which allows acceptance of the contractor's estimate of the amount of loss and/or expense he will incur in carrying out an instruction **[6.5 and 6.10]**.

The contractor's entitlement to recovery under clause 26 depends on the correct operation of its machinery. It is therefore vitally important that all those concerned – contractors, architects and quantity surveyors – should fully understand its intention and the way it works. Most of clause 26 deals with the contractor's and nominated sub-contractor's rights to financial reimbursement for events which are breaches of contract by, or which are within the control of, the employer himself, others for whom he is responsible, or the architect acting on the employer's behalf.

Some of the events to which the clause refers are not breaches of contract by the employer or those for whom he is responsible. For instance, an instruction to open up work for inspection or requiring a variation is one which the architect is specifically empowered to issue under the contract, and, therefore, its issue plainly cannot be a breach of contract which would entitle the contractor to recover damages. Therefore, clause 26 contains the only opportunity for the contractor to get compensation for such events, and if the contractor loses his right to compensation under the clause by failing to make an application at the proper time, the preservation of his common law rights under clause 26.6 will be to no avail. This consideration applies to the events described in clauses 26.2.2, 26.2.5, 26.2.7, 26.2.8, 26.2.10 and to deferment of possession of the site where clause 23.1.2 applies.

The clause is divided into six parts:

26.1   sets out the rights and obligations of the contractor if he wishes to obtain payment and the rights and obligations of the architect and/or quantity surveyor in response.

26.2   sets out the grounds that may entitle a contractor to payment.

26.3   sets out the architect's duty to notify the contractor in specific instances of the grounds upon which he has granted extensions of time. This is a misleading sub-clause **[10.13]**.

26.4   sets out the procedure for operating similar provisions in the Standard Nominated Sub-Contract Forms NSC/C under the main contract.

26.5   read with clauses 3 and 30, provides for certification and payment of amounts which the architect decides are due to the contractor.

26.6   preserves the contractor's common law and other rights. This is important in view of the Court of Appeal decision in *Lockland Builders Ltd* v. *John Kim Rickwood* (1995), which seems to suggest that, in the absence of express provision, contract machinery and common law, rights can co-exist only in circumstances where repudiation has taken place. Although that particular decision may be open to some criticism or at least be referrable to particular facts, clause 26.6 puts the matter beyond doubt.

10.3   *The claims procedure*

The general meaning of clause 26.1, which contains the meat of clause 26 is as follows:

If the contractor considers that regular progress of any part or the whole of the Works has been or is likely to be substantially affected by any of the matters in clause 26.2 and if he also considers that this or any deferment of possession may result or has resulted in direct loss and/or expense on the contract for which he would not be able to receive payment under any other term in the contract he may if he so wishes make a written application to the architect (which he may quantify), but he must do so as soon as it has become, or should reasonably have become, apparent to him.

The architect must then decide whether or not the contractor is correct and may, to assist him in forming his opinion, request the contractor to submit the information that will reasonably enable him to do so. If the architect agrees with the contractor, he must himself ascertain, or instruct the quantity surveyor to ascertain, the amount of the direct loss and/or expense which the contractor has incurred. The architect or the quantity surveyor may, in order to enable them to carry out the ascertainment, require such details of loss and/or expense as are reasonably necessary for that purpose.

10.4   *The contractor's application*

It appears that the making of a written application by the contractor at the right time is a condition precedent. Failure by the contractor to apply in writing as specified in the contract will be fatal to his claim for payment under the contract: *Hersent Offshore SA and Amsterdamse Ballast Beton-en-Waterbouw BV* v. *Burmah Oil Tankers Ltd* (1979); *Wormald Engineering Pty Ltd* v. *Resources Conservation Co International* (1992). Architects often overlook this contractual requirement and it is not unusual to find architect and contractor engaged in dialogue about a loss and/or expense claim for the first time many months after the event, indeed often months after practical completion has been certified. Although it is open to an architect to refuse to consider a later application (probably, he cannot consider it), once he starts to consider it, he has probably on behalf of the employer waived the right to reject it on these grounds.

Failure to notify the architect in advance will deprive him of the opportunity to take any remedial action open to him. An early, rather than a late, application is therefore essential to enable the contractor to demonstrate that he has taken all reasonable steps to mitigate the effect on progress and the financial consequences. If the architect fails to take advantage of this, then clearly it is his responsibility and not the contractor's when answering to the employer for the extra cost involved. The objective of the whole machinery of application is to bring the architect's attention to the possibility that disruption is likely to occur and will be costly to the employer unless he takes action to avoid it. There should be no question of the contractor's written application being made after the event if it is reasonable for him to anticipate trouble in the future. Even if it is not reasonable for the contractor to anticipate disruption, the application must be made as soon as the trouble occurs, and not just within a 'reasonable time' of it occurring.

There are occasions when the question of whether a written application under clause 26.1 should be made is itself difficult to resolve. Clearly, the making of the application should be the result of a deliberate decision made by the contractor and not an automatic response, for instance, to the issue of every architect's instruction. There are cases in which it will be difficult for the contractor to determine whether progress is likely to be affected. Circumstances may already have affected progress so that the occurrence of the new event may at the time not seem likely to have any further effect. There may also be other factors as was said in *Tersons Ltd* v. *Stevenage Development Corporation* (1963):

> 'Notice of intention to claim, however, could not well be given until the intention had been formed ... [and] it seems to me that the contractors must at least be allowed a reasonable time in which to make up their minds. Here the contractors are a limited company, and that involves that, in a matter of such importance as that raised by the present case, the relevant intention must be that of the board of management [i.e. directors] ... in determining whether a notice has been given as soon as practicable, all the relevant circumstances must be taken into consideration ... One of the circumstances to be considered in the present case is the fact that it was not easy to determine whether the engineer's orders ... did or did not involve additional work ...'

The court was there concerned with the ICE Conditions (2nd edition) and the question of whether certain notices were given 'as soon as practicable'. Similar circumstances can be envisaged in relation to architect's instructions. Contractors must make their applications under clause 26.1 at the earliest practicable time.

The application should be in writing, but no particular form is specified. Clearly it should state that the contractor has incurred or is likely to incur loss and/or expense arising directly from the deferment of giving possession of the site or the material effect upon the regular progress of the works or any part of the works of one or more of the ten matters listed in clause 26.2. The application may be sufficient if it refers to the general grounds and identifies the occurrence, stating that loss and/or expense is being or is likely to be incurred. It is much better if the application clearly specifies whether it is deferment of possession or which of the matters listed in clause 26.2 is relied upon. It is also advisable, even at this stage, for the contractor to provide as much detail as possible about the circumstances that have given rise to the application. For example, suppose the architect has postponed part of the project. The contractor's written application should refer to clause 26.2.5, and should state the date on which the instruction was issued, the work which was postponed and the effect which the postponement has had, is having and will have upon contract progress, going into as much relevant detail as possible.

The contractor need make only one written application in respect of loss and/or expense arising out of the occurrence of any one event. This will entitle him to recover past, present and future loss and/or expense arising from that event, and there is no need to make a series of applications as was the case under some former contracts. It is plain that a 'general' or 'protective' notice is not sufficient under clause 26.1. Specific written applications must be made in respect of each event. Further, the issue of an automatic standard letter application every time one of the events listed occurs does not satisfy the requirements of clause 26.1 unless it clearly refers to the appropriate grounds. Where such an application does not satisfy clause 26.1, it is invalid and submitting it is a fruitless exercise. The contractor's written application under clause 26.1 is related to the degree to which one or more of the matters listed in clause 26.2 has affected or is likely to affect regular progress, and the contractor must have sustainable grounds for believing this to be the case.

Clause 26.1.1 requires the written application to be made as soon as it has become, or should reasonably have become, apparent to him that 'regular progress of the Works or any part thereof has been or was likely to be' materially affected. In the case of deferred possession, it appears that an application should be made as soon as notification is received from the employer that possession of the site is to be deferred. The application must, therefore, be made at the earliest reasonable time and certainly before regular progress of the works is actually affected, unless there are good reasons why the contractor could not foresee that this would be the case. Although clause 26.1 allows for an application to be made at the time of or after the event, the intention is clearly that the architect should be kept informed at the earliest possible time of all matters likely to affect the progress of the work and likely to result in a claim for loss and/or expense.

10.5   The architect cannot plead ignorance of work on site

Clause 26 now clearly assumes that the architect cannot be expected to have more than a general knowledge of what is happening on site: *East Ham Corporation* v. *Bernard Sunley & Sons Ltd* (1965).

It is the contractor who is responsible for progressing the work in accordance with the requirements of the contract and the architect's instructions. The practical effect of the contractor's obligation to notify as soon as the regular progress is likely to be materially affected is quite significant: *Jennings Construction Ltd* v. *Birt* (1987). The architect is entitled to assume, unless notified to the contrary, that work is progressing smoothly and efficiently and that there are no current or anticipated problems. For instance, if he issues an instruction – even one requiring extra work – and the contractor accepts it without comment, the architect is probably entitled to assume that the effects of that instruction can be absorbed by the contractor into his working programme without any consequential delay or disruption: *Doyle Construction Ltd* v. *Carling O'Keefe Breweries of Canada* (1988). That is not invariably the case. In *London Borough of Merton* v. *Stanley Hugh Leach Ltd* Mr Justice Vinelot observed:

> 'Although I accept that the architect's contact with the site is not on a day to day basis there are many occasions when an event occurs which is sufficiently within the knowledge of the architect for him to form an opinion that the contractor has been involved in loss or expense.'

This is particularly so where there has been some default by the architect himself. It would be unsafe for an architect to rely too heavily on the strict letter of the contractor's obligation as to timing of the notice when he himself has been late in the issue of information which has been requested at the proper time and he must well know that this is bound to have a material effect on progress leading to the contractor incurring loss and expense.

10.6   *The contractor is not allowed two bites at the cherry*

The phrase 'would not be reimbursed by a payment under any other provision in this Contract' is important. It is to prevent double payment, for instance where increased costs of labour and materials during a period of delay to completion are already being recovered under the fluctuations provisions of the contract. In relation to claims arising from clause 26.2.7 (later in this chapter), obviously some care must be taken to distinguish between those costs which are covered by the quantity surveyor's valuation under clause 13 and those for which reimbursement may be obtained under this clause (see the proviso to clause 13.5). There is, however, another aspect to this phrase which is often overlooked. Contractors often claim on an 'either or' basis, hopeful that what they miss under one clause they will recover under the other. This strategy may be successful, but the use of 'would not be' rather than 'has not been' is significant. The effect is that if the contractor is entitled to be reimbursed under any other clause, he is not entitled to be reimbursed under clause 26 whether or not he has actually received reimbursement under any other clause. It seems that if he is entitled to recover under clause 13, he must persevere in his attempts for he cannot recover what amounts to a shortfall under clause 13 as loss and/or expense under clause 26.

10.7   Effect on regular progress

The contractor's entitlement to make application under clause 26 is that 'the regular progress of the Works or of any part thereof has been or is likely to be materially affected by any one or more of' the ten matters listed in clause 26.2. In other words, it is the effect of the stated event upon the regular progress of the works, i.e. any delay to or disruption of the contract progress.

Regular progress can be affected other than by delay alone. To exclude other effects pays no attention to the words used and offends against common sense and the straightforward commercial intention of the contract. There can be a disturbance to regular progress, resulting in loss of productivity in working, without there being any delay as such either in the progress or in the completion of the work. The clause cannot be interpreted so as to confine the contractor's right to reimbursement to circumstances that delay progress. It covers circumstances that may give rise, for instance, to reduced efficiency of working without progress as a whole being delayed. It should be noted, however, that this is not the same as saying that merely because the work has proved to cost more or to take longer to complete than was anticipated, the contractor is entitled to additional payment. It must be possible for him to demonstrate that the cause is directly attributable to one or more of the matters set out in clause 26.2 and the effect upon regular progress of the works. The words 'regular progress' have caused difficulty. They must be related to the contractor's obligation under clause 23.1 'regularly and diligently [to] proceed with' the works **[9.4]**. Whether there is regular progress and whether or not it has been, or is likely to be, 'materially affected' must be a matter of objective judgment in each case so far, of course, as that is possible.

The contractor's progress may be irregular, due to factors within his control or which do not give him entitlement to claim. That is not fatal to his claim under this clause although it will present severe evidential problems. Among other things, he will have to demonstrate what regular progress should have been and further prove that, irrespective of his own failures in this respect, regular progress would have been affected by the matter specified.

Regular progress must have been, or be likely to be *materially* affected. 'Materially' has been defined as, among other things, 'in a considerable or important degree'. Trivial disruptions such as are bound to occur on even the best-run contract are clearly excluded. The circumstances must be such as to affect regular progress of the works 'in a considerable or important degree'. The affectability must be of some substance, therefore 'substantially' is a more recognisable and serviceable word. The particular point at which disruption becomes considerable or important is impossible to define in general terms. It is a matter of interpretation in the light of particular circumstances.

10.8   The architect's duty

The initiative must come from the contractor, who may make a written application. He clearly has no duty to do so. Provided that the application is properly made in accordance with the terms of the contract, the initiative then passes to the architect. If the architect forms the opinion that the contractor has suffered or is likely to suffer direct loss and/or expense due to deferment of possession or because regular progress has been substantially affected as stated in the contractor's application, then as soon as he does so, he must initiate the next stage: the ascertainment of the resulting direct loss and/or expense. The architect's opinion is the key factor. Making an application does not entitle the contractor to money if, in the architect's opinion no money is due – though, of course, ultimately it may be the opinion of an arbitrator that will finally determine the matter. The process of ascertainment does not begin unless and until the architect has formed the opinion that deferment of possession has given rise to direct loss and/or expense or that regular progress has been or is likely to be materially affected as set out in the application of the contractor. If the contractor's application is not made at the proper time then the architect must reject it, whatever its merits may be, and he is not empowered to form an opinion about it [10.4]. Further, the architect can deal only with matters that are set out in the contractor's written application; he has no authority to deal with any matters affecting regular progress that are not the subject of a written application from the contractor albeit the architect may be fully aware of them and if they had been included, he would have accepted them.

When the architect has received all necessary information, he should then consider:

(a) What is the matter alleged in respect of this specific claim?
(b) Is it one where the employer (or the architect) is in some way at fault or responsible?
(c) Has the regular progress of the work been affected by the matter alleged?
(d) Has the regular progress of the work been *materially* affected?
(e) If so, has the contractor suffered direct loss and/or expense or merely consequential damage?
(f) Is it loss or expense for which he would not be reimbursed by a payment under any other provision of the contract?

10.9   What is meant by 'information'

Clause 26.1.2 entitles the architect to request the contractor to supply such further information as is reasonably necessary to enable him to form an opinion as to the effect on regular progress or whether the deferment of possession has given or is likely to give rise to direct loss and/or expense. It is in the contractor's own interest to provide as much relevant information as possible at the time of his written application and not to wait until the architect asks for it under this sub-clause. The information which the architect is entitled to request is that which should *reasonably* enable him to form an opinion; an architect is not entitled to delay matters by asking for more information than is reasonably necessary.

The architect must specify the precise information required, rather than simply requiring the contractor to 'prove' his point. The contractor is entitled to know what would satisfy the architect and enable him to form a view. Whether or not this is the position in law, it certainly should be the aim of the architect, who might otherwise be accused of delaying tactics. It is suggested that such details might include comparative programme/progress charts in network form pin-pointing the effect upon progress, together with the relevant extracts from wage sheets, invoices for plant hire, etc. Mr Justice Vinelot in *London Borough of Merton* v. *Stanley Hugh Leach Ltd* (1985) said:

'If [the contractor] makes a claim but fails to do so with sufficient
particularity to enable the architect to perform his duty or if he
fails to answer a reasonable request for further information he
may lose any right to recover loss or expense under those sub-
clauses and may not be in a position to complain that the architect
was in breach of his duty.'

These are sensible words which highlight not only the contractor's
responsibility, but also the consequences if he refuses to help himself. It is important to consider the other side of the coin. The
architect's or the quantity surveyor's requests for further information must be reasonably precise. When he receives the request, the
contractor should be able to say with a fair degree of accuracy what
he must provide. Endless vague requests, as a delaying tactic, are all
too common.

The duty of the architect or the quantity surveyor is to ascertain
the amount of the direct loss and/or expense as a matter of fact and
it is necessary for them to look to the contractor as the person in
possession of the facts to provide them. Clause 26.1.3 provides that
the contractor must submit on request such details of the loss and/
or expense as are reasonably necessary for the ascertainment. This
does not necessarily mean the submission of a calculated claim,
although it may well be in the contractor's interest to provide it,
particularly if it is expected that the matter may move to arbitration.
Clause 26.2.1 now somewhat unnecessarily says that the contractor
may submit a calculated claim.

As a basic rule, the contractor should be requested to provide no
more than is strictly necessary, indeed clause 26.1.3 states as much,
and the necessary information must be particularised by the
architect or the quantity surveyor. On receipt of the request the
contractor should know that when it is provided, ascertainment of
the whole claim can be completed without delay.

10.10   The ascertainment

The word 'ascertainment' means 'finding out for certain'. It is not
therefore simply a matter for the judgment of the architect or
quantity surveyor; the duty upon them is *to find out* the amount of
the direct loss and/or expense for certain, not to estimate it. It also
follows that the loss and/or expense that has to be found out must
be that which is being, or has been actually incurred: *Alfred
McAlpine Homes North Ltd* v. *Property & Land Contractors Ltd* (1995).

References to estimated figures included in the contract bills usually have no relevance. However it must always be borne in mind that the architect cannot refuse to certify payment to the contractor of a reasonable assessment of direct loss and/or expense that he has incurred because no better information is available, although in such circumstances the contractor may have to accept an assessment figure that is less than his actual loss. The architect may instruct the quantity surveyor to ascertain the direct loss and/or expense. It has been said that the architect is then bound by the quantity surveyor's ascertainment. Although this may be the practical effect of passing the duty to ascertain to a quantity surveyor, responsibility for certification of the amount still lies with the architect who may be held to be negligent if he was to certify without taking reasonable steps to satisfy himself of the correctness of the amount: *Sutcliffe* v. *Thackrah* (1974). There is nothing in the contract which suggests that the architect is bound to accept the quantity surveyor's opinion or valuation when he exercises his own function of certifying sums for payment: *R.B. Burden Ltd* v. *Swansea Corporation* (1957). It is essential, however, that the architect's instruction or his request for assistance be precisely set out in writing. The quantity surveyor's agreement to assist must also be in writing so as to establish the quantity surveyor's responsibility to the architect should his advice be given negligently.

10.11   The matters

Judge Edgar Fay said in *Henry Boot Construction Ltd* v. *Central Lancashire New Town Development Corporation* (1980) of JCT 63 provisions equivalent to those now found in JCT 98, clauses 25 and 26:

'The broad scheme of these provisions is plain. There are cases where the loss should be shared, and there are cases where it should be wholly borne by the employer. There are also those cases which do not fall within either of these conditions and which are the fault of the contractor, where the loss of both parties is wholly borne by the contractor. But in the cases where the fault is not that of the contractor the scheme clearly is that in certain cases the loss is to be shared; the loss lies where it falls. But in other cases the employer has to compensate the contractor in respect of the delay, and that category, where the employer has to compensate the contractor, should, one would think, clearly be composed of cases where there is fault upon the employer or fault for which the employer can be said to bear some responsibility.'

This is still worthy of careful study. Clause 26.2 deals with those cases where the employer must compensate the contractor and it is only if the contractor can establish that he has incurred direct loss and/or expense not otherwise reimbursable as a direct result of one or more of the nine matters listed in the clause (or, of course, as a direct result of deferment of possession of the site) that he is entitled to reimbursement. Each of these seven matters are discussed below.

10.11.1   Late information: clause 26.2.1

This ground covers failure by the architect to provide necessary information to the contractor at the proper time to enable the contractor to use it for the purpose of the works. Since the revision of clause 5.4 into clauses 5.4.1 and 5.4.2 **[2.11]**, this ground has been considerably simplified. Essentially, the contractor has to show that the completion date is likely to be delayed, because the architect failed to comply with his duty, either to comply with the information release schedule or with his general obligation to provide information as set out in clause 5.4.2.

The wording of this ground is identical to clause 25.4.6 and reference should be made to the commentary on that clause **[7.11.2]**.

The contractor has a date to work to; initially it is the date stated in the appendix and, later, if the architect makes an extension of contract time, it is the new date so fixed. The contractor has an obligation to work towards that date and he is also entitled not to be hindered in his efforts to achieve that date by any act or default on the part of the architect or the employer.

10.11.2   Opening up for inspection and testing: clause 26.2.2

The architect is empowered under clause 8.3 to require work to be opened up for inspection and to instruct the contractor to arrange for or to carry out the testing of materials and executed work in order to ensure that they comply with the contract. It is a curious feature of the wording of clause 8.3 that the cost of such opening up and testing is to be reimbursed to the contractor unless already provided for in the contract documents or unless the inspection or tests show that the materials or work are not in accordance with the contract. In other words it is for the architect to show that they are not in accordance with the contract and not for the contractor to show that they are. The same considerations apply to the question of whether the contractor is entitled to an extension of time under clause 25.4.5.2 **[7.11.1]** and to the question of entitlement to reimbursement under this clause. Each of these provisions assumes that the contractor will be entitled unless the inspection or tests demonstrate that the materials or work are not in accordance with the contract. The burden of proof is probably reversed if the contractor has prematurely covered up work when the preliminaries direct that it must be left open until it has been inspected by the architect.

10.11.3   Any discrepancy in or divergence between the contract drawings and/or the contract bills and/or the numbered documents: clause 26.2.3

The simplicity of the wording of clause 26.2.3 contrasts sharply with the elaborate provisions of clause 2.3, which refers to discrepancies in or divergences between any two or more of the following:

(1)   the contract drawings;
(2)   the contract bills;
(3)   any architect's instructions other than variations;
(4)   any necessary drawings or documents issued by the architect for the general purposes of the contract;
(5)   the numbered documents.

No doubt it is considered that these other discrepancies are adequately covered by other entitlements. The contractor's obligation, however, is not to find discrepancies, but merely to notify the architect if he does find them: *London Borough of Merton* v. *Stanley Hugh Leach Ltd* (1985). Therefore, he may not discover a discrepancy until after that portion of the work has been constructed. That does not prevent the contractor from claiming on this ground. Whether he is successful will depend on all the circumstances, not least whether he has complied with the clause 26 procedures and, perhaps, whether he should have been aware (and, therefore, reported) the discrepancy.

10.11.4   Work not forming part of the contract: clause 26.2.4

Clause 26.2.4 covers the situation where the employer himself carries out work not forming part of the contract or arranges for such work to be done by others at the same time as the contract works are being executed by the contractor; it also covers the situation where the employer has undertaken to provide materials or goods for the purposes of the works. Clause 29 provides that, where the work in question is adequately described in the contract bills, so that the contractor has been able to make adequate allowance for the effect on him in the contract sum and in his programme for the works, the contractor must permit such work to be done [4.16]. It will only be if the work concerned causes an unforeseeable delay or disruption to the contractor's own work that any claim will lie under this sub-clause. However, clause 29 also provides that if the work in question is not adequately described in the contract bills but the employer wishes to have such work executed, the employer may arrange for it to be done with the consent of the contractor, which is not to be unreasonably withheld. In that event, a claim will almost inevitably arise under clause 26.2.4.1 since the contractor will not have been able to make adequate allowance in his programme and price.

A clause 26.2.4.1 claim may sometimes relate to work carried out by statutory undertakers, if the work is carried out by them as a matter of contractual obligation, rather than as one of statutory obligation. When acting under statutory obligations, any interference on their part with the contractor's work does not give rise to any monetary claim against the employer under this or any other provision in the contract, but may give rise to a claim for extension of time under clause 25.4.11. An arbitrator found as a fact in regard to a particular project that the statutory undertakers 'were engaged under contract by the respondents to construct the said mains', and it would appear that he also found that most, if not all, of the work being carried out by the statutory undertakers was not covered by their statutory obligation, but was being executed by them voluntarily at the employer's request and expense. At the trial of *Henry Boot Construction Ltd* v. *Central Lancashire New Town Development Corporation* (1980), Judge Edgar Fay said:

'These statutory undertakers carried out their work in pursuance of a contract with the employers; that is a fact found by the arbitrator and binding on me. ... In carrying out [their] statutory obligations they no doubt have statutory rights of entry and the like. But here they were not doing the work because statute obliged them to; they were doing it because they had contracted with the [employers] to do it.'

Two points of interest with regard to clause 26.2.4:

(a) the ground on which the contractor may apply is not simply the employer's *failure* to carry out the work or supply materials. He may claim if the employer executes the work or supplies the materials perfectly properly and at the right time. This lays a considerable burden on the employer and makes the employer's decision to employ others on the Works something akin to writing a blank cheque.

(b) although clause 26.2.4.2 refers to supply or failure to supply materials which the employer has agreed to provide for the Works, the contract does not make provision for this agreement to take place. It is not included within clause 29, although perhaps it should be. Therefore, any agreement of this kind must take place outside the building contract and it should be in writing.

10.11.5   Postponement of work: clause 26.2.5

Clause 26.2.5 refers to the power of the architect, under clause 23.2, to 'issue instructions in regard to the postponement of *any work* to be executed under the provisions of (the) Contract' (author's emphasis). Clause 23.2 does not empower the architect to issue an instruction postponing the date upon which possession of the site is to be given to the contractor as stated in the appendix, nor does he have power to do so under any other express or implied term of the contract, and this is emphasised by clause 23.1.2 which gives the employer, and not the architect, the right to defer possession of the site for a period not exceeding six weeks or any shorter period stated in the appendix **[7.8]**. Otherwise the ground is clear enough.

Although the clause refers to instructions issued under clause 23.2, such instructions may arise as a matter of fact: *M. Harrison & Co. (Leeds) Ltd* v. *Leeds City Council* (1980).

10.11.6   Failure to give ingress to or egress from the site: clause 26.2.6

This clause deals with the situation if either there is a provision in the contract bills and/or drawings, or an agreement has been reached between the contractor and the architect, permitting the contractor means of access to the site 'through or over any land, buildings, way or passage adjoining or connected with the site and *in the possession and control of the Employer*' (author's emphasis) and the employer fails to keep his side of the agreement.

The words emphasised above should be noted; the clause does not apply where the employer may have undertaken to obtain a wayleave across land which is not in his possession and control. In such a case failure by the employer to obtain the wayleave would give rise to a claim at common law. It should be further noted that the land, etc. must be in both the possession *and* the control of the employer, and the land, etc. must also be 'adjoining or connected with the site', which would suggest that there need not necessarily be physical contact.

10.11.7 Variations and work against provisional sums: clause 26.2.7

Variations and instructions issued by the architect for the expenditure of provisional sums are dealt with in clause 13, which is discussed in Chapter 3. Clause 26.2.7 covers disturbance costs where the introduction of the variation or provisional sum work materially affects the regular progress of the works in general. Instructions for the expenditure of provisional sums for performance specified work are excluded, because the contractor will already know the scope and extent of such work and he will have been able to allow for associated costs at the time of tender. It must be noted that the contractor's priced statement and the clause 13A quotation each provides for loss and/or expense to be part of the contractor's overall estimate for carrying out the variation.

10.11.8 Approximate quantity not a reasonably accurate forecast of the quantity of work: clause 26.2.8

This clause is intended to cover the situation where an approximate quantity has been included in the bills of quantity, but the quantity of work actually executed under that item is different; either greater or less. As long as the approximate quantity is reasonably accurate, the contractor has no claim. What is 'reasonably accurate' will depend on all the circumstances, but as a rule of thumb an approximate quantity which was within 10% of the actual quantity would be difficult to demonstrate as unreasonable. The contractor's entitlement will usually be based on the extra time he requires over and above the time he has allowed for doing the quantity of work in the bills. The approximate quantity may be an unreasonable estimate, because the actual quantity is substantially less than in the bills. Theoretically, the contractor will also have grounds for a claim under this head, but it will take considerable ingenuity to put together. It should be noted that this clause expressly refers to 'work'. The conclusion is that increases in materials will not entitle the contractor to claim. Generally, it is only an increase in work or labour which will require extra time to execute, but there may be circumstances where increases in the quantity of materials may result in additional off-site and unquantified work, such as in the drawing office or the fabrication shed. It appears that the contractor will have no claim for such matters, at least under this head.

10.11.9   Compliance or non-compliance with duties in relation to the CDM Regulations: clause 26.2.9

Clause 6A provides, among other things, that the employer will ensure that the planning supervisor carries out all his duties under the CDM Regulations and, where the contractor is not the principal contractor for the purpose of the Regulations, that the principal contractor carries out all his duties under the CDM Regulations. It will be usual for the main contractor to be the principal contractor for the purposes of the Regulations and the employer's duty will just relate to the planning supervisor. The obligation placed upon the employer to 'ensure' is virtually to guarantee that the planning supervisor will carry out his duties. That has been the view of the court where a party has an obligation to 'ensure' or 'secure' the doing of something: *John Mowlem & Co Ltd* v. *Eagle Star Insurance Co Ltd* (1995) confirming the judgment of the Official Referee. The court made clear their view that to 'ensure' meant exactly what it said and amounted to more than a mere obligation to use best endeavours. It should be noted that the contractor's entitlement to recover loss and/or expense does not depend on the employer's failure to comply with his obligations under clause 6A. Compliance can also be a ground, provided of course that the other conditions are satisfied. This is important, because compliance with the CDM Regulations may involve the contractor in unexpected time and expense when an architect's instruction is involved.

10.11.10   Clause 30.1.4 suspension: clause 26.2.10

Clause 26.2.10 introduces this ground although it is not required under the Housing Grants, Construction and Regeneration Act 1996 which entitles a contractor to suspend performance of his obligations on seven days written notice if the employer does not pay a sum due 'in full by the final date for payment'. The suspension part of section 112 is dealt with by clause 30.1.4 and it should be noted that copying the notice to the architect, while not required by the Act, appears to be a condition precedent to the contractor's entitlement to loss and/or expense under the contract **[9.15]**.

The Act rather unrealistically assumes that the contractor will be satisfied with a bare extension of time. Of course, the contractor may suffer severe financial hardship due to unlawful withholding of money properly due and this is recognised in clause 26.

10.12 *Deferment: clause 26.1*

Before the introduction of clause 23.1.2, the contractor's right to possession of the site on the date given in the appendix was an absolute one. Possession refers to the whole of the site and, in the absence of sectional possession, the employer is not entitled to give possession in parcels: *Whittal Builders* v. *Chester Le Street District Council* (1987). It has been suggested that provided 'the contractor has sufficient possession, in all the circumstances, to enable him to perform, the employer will not be in breach of contract'. However, his right to possession is an express term of the contract (clause 23.1) and in any event there is at common law an implied term in any construction contract that the employer will give possession of the site to the contractor in time to enable him to carry out and complete the work by the contractual date: *Freeman & Son* v. *Hensler* (1900). In *London Borough of Hounslow* v. *Twickenham Garden Developments Ltd* (1970), Mr Justice Megarry said 'The contract necessarily requires the building owner to give the contractor such possession, occupation or use as is necessary to enable him to perform the contract'.

Accordingly, subject to the right to defer possession for up to six weeks under clause 23.1.2 if that clause is stated in the contract appendix to apply, any failure by the employer to give possession on the due date is a breach of contract, entitling the contractor to bring a claim for damages at common law in respect of any loss that he suffers as a consequence: *London Borough of Hounslow* v. *Twickenham Garden Developments Ltd* (1970).

Moreover, if clause 23.1.2 does not apply and the employer fails to give possession of the site to the contractor on the due date, or if the clause does apply and possession is deferred for more than the stated maximum period, since the architect has no power to grant an extension of time for completion on that ground it follows that the employer will forfeit any right to liquidated damages and the contractor's obligation will be to complete within a reasonable time: *Wells* v. *Army & Navy Co-operative Society Ltd* (1902); *Amalgamated Building Contractors Ltd* v. *Waltham Holy Cross UDC* (1952); *Peak Construction (Liverpool) Ltd* v. *McKinney Foundations Ltd* (1970). The architect has no power to extend time on grounds other than those set out in clause 25: *Percy Bilton Ltd* v. *Greater London Council* (1982).

10.13   *The curious clause 26.3*

If ever there was a superfluous clause and, moreover, one which apparently flies in the face of common sense and case law, it is clause 26.3. This clause provides that if and to the extent that it is necessary for the purpose of ascertainment of direct loss and/or expense, the architect shall state in writing to the contractor what extension of time, if any, he has granted under clause 25 in respect of those events which are also grounds for reimbursement under clause 26.

There is no logical justification for the inclusion of this provision, which appears to give contractual sanction to the mistaken belief that there is some automatic connection between the granting of an extension of time and the contractor's entitlement to reimbursement. There is or ought to be no such connection: *H. Fairweather & Co Ltd* v. *London Borough of Wandsworth* (1987). An extension of time under clause 25 has only one effect. It fixes a new date for completion and, in so doing, it defers the date from which the contractor becomes liable to pay to the employer liquidated damages. An extension of contract time does not in itself entitle the contractor to any extra payment. JCT 98 clause 25 entitles the contractor to relief from paying liquidated damages at the date named in the contract. He is certainly not entitled to claim items set out in 'Preliminaries' for the extended period. Moreover, this information is of no interest or relevance to the contractor and cannot and should not have any relevance to any ascertainment of direct loss and/or expense under clause 26. It is for the architect or quantity surveyor to ascertain the amount of loss and/or expense. Clause 26.3 apparently merely requires the architect to specify the relevant events he has taken into account without apportioning. The clue is in the words 'If and to the extent it is *necessary for ascertainment*' (author's emphasis). The judge in *Methodist Homes Housing Association Ltd* v. *Messrs Scott & McIntosh* (1997) found it difficult to think of any situation when the statement of extension of time made under clause 25 would be at all necessary for the purpose of ascertainment.

10.14   *Nominated sub-contract claims*

Clause 4.38 of JCT Nominated Sub-Contract Form (NSC/C) is a provision corresponding to clause 26 of the main contract form enabling the nominated sub-contractor to claim against the employer *through* the main contractor in respect of direct loss and/ or expense and on similar grounds to those given to the main contractor by clause 26. Clause 26.4 provides the necessary machinery by which the contractor is to pass on such claims to the architect, and the architect's decision is to be passed back to the nominated sub-contractor.

Clause 26.4.2 corresponds to clause 26.3 in a nominated sub-contractor situation and the same objections are equally sustainable in respect of it **[10.13]**.

10.15   *Certification timing*

Clause 26.5 provides for amounts ascertained to be added to the contract sum. By clause 3, as soon as an amount of entitlement is ascertained in whole or in part, that amount is to be taken into account in the next interim certificate. By clause 30.2.2.2, such amounts are not subject to retention. The reference to ascertainment 'in part' in clause 3 means that it is not necessary for the full process of ascertainment to have been completed before an amount must be certified for payment. This is important from the point of view of both the contractor and the employer: in particular, where the direct loss and/or expense is being incurred over a period of time. The proper operation of the contractual machinery should ensure that the matter is dealt with in interim payments from month to month so far as is practicable. This provision for payment of sums ascertained 'in part' requires the inclusion in interim certificates of allowances for direct loss and/or expense as soon as some or even some part of the ascertainment has been made, thereby ensuring proper cash-flow to the contractor and reducing the employer's possible liability for financing charges: *F.G. Minter Ltd* v. *Welsh Health Technical Services Organisation* (1980).

10.16   *Other rights and remedies*

Clause 26.6 preserves the contractor's common law and other rights. The Court of Appeal decision in *Lockland Builders Ltd* v. *John*

Kim Rickwood (1995), suggested that where a party's common law rights were not expressly reserved, they could co-exist with the contractual machinery only where the other party displayed a clear intention not to be bound by the contract. This view may be open to question and seems to ignore earlier contrary authority: *Modern Engineering (Bristol) Ltd* v. *Gilbert Ash (Northern) Ltd* (1974); *Architectural Installation Services Ltd* v. *James Gibbons Windows Ltd* (1989). Possibly it will ultimately be held to be a decision based on its own particular facts. The rights set out in clause 26 confer a specific contractual remedy on the contractor in the circumstances there defined, and subject to the conditions imposed by the contract. But the contractor's other rights are expressly stated to be unaffected, in particular his right to claim damages for breach of contract: *London Borough of Merton* v. *Stanley Hugh Leach Ltd* (1985).

The specific contractual right to claim reimbursement for direct loss and/or expense is additional to any rights or remedies which the contractor possesses at law, and notably to damages for breach of contract. It may be that because of the limitations imposed by the contract machinery the contractor may decide to pursue these independent remedies. However, some events are not breaches of contract **[10.2]**.

10.17   *Loss and/or expense following the discovery of antiquities*

Clause 34 provides for what is to happen if 'fossils, antiquities and other objects of interest or value' are found on site or during excavation. The contractor must use his best endeavours not to disturb the object. He must cease work as far as is necessary and take all steps necessary to preserve the object in its position and condition. He must inform the architect or the clerk of works. The architect is then required to issue instructions and the contractor may be required to allow a third party, such as an expert archaeologist, to examine, excavate and remove the object. All this will almost undoubtedly involve the contractor in direct loss and/or expense and clause 34.3.1 provides that this is to be ascertained by the architect or quantity surveyor without necessity for further application by the contractor, although he will make an application in order to protect his position.

There are no provisions similar to those in clauses 26.1.2 and 26.1.3 requiring the contractor to give further information and details. In practice, an ascertainment can only be made if the relevant information and details are provided by the contractor to the

architect or quantity surveyor and he will be wise to do so. Clause
34.3.2 is open to similar objections to those already made in respect
of clause 26.3.

10.18   Claims for site conditions

The general law is that the employer does not warrant to a contractor that the site is fit for the Works designed for it or that the
contractor will be able to construct the building on the site to that
design: *Appleby* v. *Myers* (1867). If, therefore, the contractor meets
difficulties which he did not anticipate, he is not entitled to extra
payment for them: *Bottoms* v. *York Corporation* (1892).

There are two main kinds of claim under this category:

- If the contractor is given incorrect information about site conditions
- In relation to the particular provisions of JCT 98, where clause
  2.2.2.1 provides that the bills are to have been prepared in
  accordance with the Standard Method of Measurement, 7th
  edition.

The contractor may have a claim for negligent misrepresentation
and/or breach of warranty and/or under the Misrepresentation Act
1967, as amended, arising from misrepresentations made by or on
behalf of the employer. As a result of the Misrepresentation Act
1967, the remedies which were formerly restricted to cases of fraud
or recklessness apply to all misrepresentations unless the party who
made the representation can prove 'that he had reasonable ground
to believe and did believe up to the time the contract was made that
the facts represented were true'.

Liability for misrepresentation is not affected by the old rule that
the employer does not warrant that the site is fit for the Works or
that the contractor will be able to construct on the site: *Appleby* v.
Myers (1867). Architects will be personally liable at common law for
any fraudulent or negligent misstatement or representation and
also under the 1967 Act.

In an appropriate case, the contractor may have a claim against
the employer for misrepresentations about site and allied conditions made during pre-contractual negotiations. (Indeed, the
architect can become personally liable: *J. Jarvis and Sons Ltd* v. *Castle
Wharf Developments & Others* (2001).)

In the Australian case of *Morrison-Knudsen International Co Inc* v.

Commonwealth of Australia, the contractor claimed that basic information provided to him at pre-tender stage was false, inaccurate and misleading. On a preliminary issue, it was concluded:

> 'The basic information in the site document appears to have been the result of much technical effort on the part of a department of the defendant. It was information which the plaintiffs had neither the time nor the opportunity to obtain for themselves. It might even be doubted whether they could be expected to obtain it by their own efforts as a ... tenderer. But it was indispensable information if a judgment were to be formed as to the extent of the work to be done...'

In *Holland Hannen & Cubitts (Northern) Ltd* v. *Welsh Health Technical Services Organisation* (1981), one of the claims made by the contractors against the employers was for damages for negligent misrepresentations and/or breach of warranty and/or pursuant to the Misrepresentation Act 1967 arising out of representations made or warranties given by or on behalf of the employer. These related, among other things, to statements in the preliminaries section of the bills of quantities about the sequence of operations, letters from the architects, and statements made at pre-contractual meetings. In an appropriate case, therefore, an action would be possible for a misleading statement about site conditions.

The second possibility is in relation to clause 2.2.2.1 and the obligation to comply with SMM 7. This seems to require the employer to provide the contractor with information in his possession about potentially difficult site conditions. Other provisions require the employer to provide specific information. The contractor may have a claim against the employer, should site conditions *not* be as assumed: *C. Bryant & Son Ltd* v. *Birmingham Hospital Saturday Fund* (1938). SMM 7 states that information regarding trial pits or bore holes is to be shown on location drawings under 'A. Preliminaries/General Conditions' or on further drawings which accompany the bills of quantities or stated as assumed. Rock is classified separately.

CHAPTER ELEVEN
DISPUTE RESOLUTION

11.1 Adjudication

In compliance with the Housing Grants, Construction and Regeneration Act 1996 (commonly known as the 'Construction Act') JCT 98 in common with other JCT contracts contains an adjudication clause (41A).

It applies where either party requires a dispute or difference arising under the contract to be referred to adjudication. Although rougher than arbitration, it is steadily increasing in popularity because of the speed with which it is conducted. Given the correct circumstances, a contractor deprived of money can recover it very quickly. It has two perceived disadvantages. The first is that it is not final and the disgruntled party may take the whole matter to arbitration or litigation dependent on the option chosen in the contract **[11.2]**. The second is that, unlike arbitration, the successful party cannot usually recover its costs unless both parties agree to give the adjudicator that power: *Northern Developments (Cumbria) Ltd* v. *J. & J. Nichol* (2000). This second point is seen by some as an advantage in that it discourages a cash rich party from spending a lot of money on the preparation of its case and it encourages parties to do it themselves. However, parties who do it themselves without proper assistance in complicated matters may involve the adjudicator in much additional work which will necessarily be reflected in his fee.

There are some contractors who have rewritten the adjudication procedures in sub-contracts so that the party initiating the adjudication is responsible for all costs, including the adjudicator's fees, win or lose. Unbelievably, provisions to this effect were held to be valid in *Bridgeway Construction Ltd* v. *Tolent Construction Ltd* (2000). It appears that the judge was not referred to section 13 of the Unfair Contract Terms Act 1977 which clearly invalidates such provisions.

The courts are showing resolution in upholding adjudicators' decisions and refusing to grant as a stay of execution, even where the other party is actively pursuing arbitration or other proceedings: *Macob Civil Engineering Ltd* v. *Morrison Construction Ltd* (1999). That principle was upheld in the Court of Appeal even where the adjudicator's decision was obviously and disastrously wrong: *Bouygues UK Ltd* v. *Dahl-Jensen UK Ltd* (2000).

11.2 *Appointment of adjudicator*

Clause 41A.2 states that the adjudicator must be either an agreed individual or someone nominated by the nominator named in the appendix. The standard nominators in the appendix are the President or Vice-President, Chairman or Vice-Chairman of either the Royal Institute of British Architects, the Royal Institution of Chartered Surveyors, the Construction Confederation or the National Specialist Contractor Council. If the person drawing up the contract for execution by the parties forgets to 'Delete all but one' of the organisations, the default organisation is the Royal Institute of British Architects. There is nothing to prevent the parties writing in another nominating body of their own choosing although no special provision has been made for that.

There is no provision for the parties to write in the name of an adjudicator. There are sound reasons for not doing so. When a dispute arises for referral to the adjudicator, the adjudicator may be ill or on holiday or otherwise unable to take the appointment. Worse still, he may be able to act but unsuitable for the particular dispute. On the other hand, it is thought by some that there is a considerable benefit in an adjudicator not only being named in the contract, but in being kept fully informed of the progress of the project throughout, so that in the event of a dispute it can be referred and decided very quickly.

The appointment of an adjudicator should be accomplished within seven days of the date of the notice of intention to refer to dispute. In practice, whether the parties are to agree or a nominating body is asked to appoint, there is seldom a good reason why the adjudicator cannot be appointed within two days. It is a term of the contract (clause 41A.2.1) that an adjudicator must not be agreed or appointed if he will not complete the JCT Standard Agreement.

If an adjudicator for some good reason cannot continue the adjudication, the appointment process is effectively restarted (clause 41A.3).

Clause 41A.8, following the Act, importantly provides that the adjudicator, his employee or agent is not liable for anything he does or fails to do while carrying out his function as adjudicator. This immunity disappears if he acts in bad faith. An obvious case would be if the adjudicator had an existing commercial relationship with one of the parties, did not divulge it and made his decision to favour that party.

11.3   Notice of intention: importance of contents

The adjudication process is started by the giving of a notice of intention to refer by one party to the other. Clause 41A.4.1 states that the notice should briefly identify the dispute or difference. By the time the notice is given, the referring party will know precisely what it wants the adjudicator to decide, indeed by that time it is not unusual for the whole of the referral to have been completed, sometimes at great length and contrary to the spirit of the adjudication process. Care should be taken with the notice, because it is the description of the dispute in the notice which will determine what the adjudicator may decide: *Fastrack Contractors Ltd* v. *Morrison Construction Ltd* (2000). In essence, he is not entitled to answer a question which has not been put to him, unless it is a question which it is necessary for him to answer in order to answer the question actually asked: *Karl Construction (Scotland) Ltd* v. *Sweeney Civil Engineering (Scotland) Ltd* (2001).

11.4   Referral

Clause 41A.4.1 also deals with the referral document. The referral is the claim. It must contain details of the dispute together with the referring party's case and any evidence which the adjudicator is asked to consider. Obviously, it must also say what it wants the adjudicator to do – award as a payment, interest, etc. Copies are sent to the adjudicator and to the other party.

The other party (often called the 'respondent') is given seven days to respond to the referral. If, as sometimes happens, the referral consists of three thick files, the task of absorbing the case, let alone answering it, is almost impossible. Any adjudicator who is the recipient of three thick files of referral knows that the referring party has carried out an 'ambush', so termed for obvious reasons. Most adjudicators will do their best to see that the referring party does not get an unfair advantage by this tactic.

The documents should be delivered by actual, special or recorded delivery. They may be delivered by fax, but clause 41A.4.2 requires delivery of hard copy thereafter for record purposes.

The adjudicator has 28 days in which to come to a decision (clause 41A.5.3), but to comply with the Act, the referring party may consent to a further 14 days extension on this period. If both parties agree, which seems unlikely given the circumstances, the period may be extended by any amount.

Clause 41A.5.7 makes clear that the parties must pay their own costs. It has been suggested that the parties can jointly give the adjudicator power to award costs, but he could not have that power without the agreement of both parties, especially in the face of this clause. However, a joint agreement may be inferred if both parties separately request their costs to be awarded against the other.

11.5 Adjudicator's powers

The adjudicator's powers are set out in clause 41A.5.5. Crucially, he must act impartially, but he may set his own procedure and take the initiative in ascertaining the facts and the law. It has been suggested that he is not bound by the full rules of natural justice (*Straume (UK) Ltd* v. *Bradlor Developments* (1999)), but the adjudicator's decision may be invalidated if he is not scrupulous in notifying each party about contacts with the other: *Discain Project Services Ltd* v. *Opecprime Developments Ltd* (2001).

The adjudicator may use his own knowledge and experience, order opening up and testing, require any further information, visit the site, obtain information or advice from others after prior warning and give a cost estimate and determining interest payments in view of the contract provisions.

The parties are jointly and severally liable for the adjudicator's fees (clause 41A.6.2), but he may direct how his fees are to be paid. If he fails to say who is to pay, the parties must each pay half (clause 41A.6.1).

11.6 A question of jurisdiction

Clause 41A.7.1 states that the adjudicator's decision is binding until the dispute is finally determined by arbitration, by legal proceedings or by agreement in writing. In practice any of those is unlikely. Most disputes end with the adjudicator's decision and compliance by one of the parties (clause 41A.7.2). However, if a party fails to comply, the other may take legal action to ensure compliance (clause 41A.7.3).

It has been stated earlier **[11.1]** that the courts almost invariably uphold the adjudicator's decision, right or wrong. The decision will not be upheld if the adjudicator has acted in excess of his jurisdiction: *Homer Burgess Ltd* v. *Chirex (Annan) Ltd* (2000); for example, where the contract was not concerned with 'construction operations'. An interesting question concerns whether in such an instance the adjudicator would still benefit from the exclusion of liability. Probably he would benefit, because clause 41A.8 refers to him acting in the discharge 'or purported discharge' of his functions. It seems, therefore, that provided he intended to act correctly, he would be protected. In some instances, the adjudicator's decision may be a nullity, because he has failed to exercise his jurisdiction: *Ballast plc* v. *The Burrell Company (Construction Management) Ltd* (2001).

11.7 *Arbitration*

JCT 98 imposes no restrictions as to when certain matters may be referred to arbitration and under it arbitration can take place on any matter at any time.

Arbitrators appointed under a JCT arbitration agreement are given extremely wide express powers. Their jurisdiction is to decide any dispute or difference arising under the contract or connected with it (article 7A). That general authority is extensive (*Ashville Investments Ltd* v. *Elmer Contractors Ltd* (1987)) and by clause 41B.2 extends to:

- Rectification of the contract
- Directing the taking of measurements or the undertaking of such valuations as he thinks appropriate
- Ascertaining and awarding any sum that he considers ought to have been included in any payment
- Opening up, reviewing and revising any account, opinion, decision, requirement or notice issued, given or made and to determine all matters in dispute as if no such account, opinion, decision, requirement or notice had been issued, given or made.

The Construction Industry Model Arbitration Rules (CIMAR), 1998 edition and current at the contractual base date, govern the proceedings (clause 41B.6). Those rules, coupled with the extensive revisions to the arbitration provisions in clause 41B of the contract now amount to a fundamental overhauling of the arbitration process necessary to bring it into line with the new 1996 Arbitration Act.

Provisions relating to arbitration now first appear in article 7A. Subject to the exercise of any prior right to have the issue initially adjudicated, all disputes or differences arising under or connected with the contract and arising between contractor and employer, or the architect on his behalf, 'shall' be referred to arbitration. If either party, mistakenly or otherwise, attempts to bypass the agreed route to arbitration and instead begins proceedings in the courts they will very soon come unstuck. There are three exceptions to that position and the following matters are specifically excluded from the arbitral process:

- Disputes about Value Added Tax
- Disputes under the construction industry scheme (CIS), provided statute dictates some other method of resolving the dispute
- Matters in connection with the 'enforcement' of any decision of an adjudicator.

Where the issue is one that falls under one or other of the first two exceptions, the appropriate statutory tribunal, in the former case the Commissioners for Customs and Excise, will have authority to hear and decide the matter. Only when the question is one concerning non-compliance with any decision previously made by an adjudicator will the parties be free to begin legal proceedings to secure such compliance. In such cases, the signs now are that the courts will take a robust view of the parties' obligation to conform with any such decision. Even then however, the courts will still only play an interim role. They may make an order regarding the enforcement of the adjudicator's decision but only in so far as that order will be made pending a final determination in arbitration of the adjudicated matter: *Macob Civil Engineering* v. *Morrison Construction* (1999).

With the incorporation of clauses 41B.4 and 41B.4.1 into their agreement, the employer and contractor agree, pursuant to section 45 of the Act, that either party may by proper notice to the other and to the arbitrator apply to the courts to determine any question of law arising in the course of the reference. Although not now the first and only available means of formal dispute resolution, arbitration, if chosen in preference to litigation, will remain for all material purposes the last resort. As such the parties and architect alike should do all they can to avoid it. It is like marriage, it should not be entered into lightly or unadvisedly.

Arbitration can be a costly, time consuming and inevitably risky venture. The eventual outcome is always uncertain and those involved in the contract should do everything possible to avoid it. Some contractors nevertheless will threaten arbitration over trivial matters in an attempt to persuade the architect to alter a decision which they dislike. Others and similar minded employers alike may use the risk, however small, of a successful outcome with an associated award for costs in their favour to force an offer of settlement against what might otherwise be considered merely a speculative and unmeritorious claim or set-off. Unfortunately, even with the recent review of dispute resolution procedures and introduction of the adjudication process, that approach is unlikely to disappear overnight. Wise contract administrators must therefore deal with speculative threats of arbitration firmly. Despite even the most strenuous efforts to do so, it will not always be possible to avoid arbitration and so employers and contractors must ensure that they properly appreciate how the process operates. Only then can they recognise the possible consequences of embarking on formal proceedings, both in terms of time and cost.

It is commonly misunderstood that the arbitration process is nothing than more or less an airing of each party's opinions and arguments in a semi-formal debate during which each party simply argues out their position, on a rather ad hoc basis, before the arbitrator then decides, in a rather casual manner, whose story he prefers. Indeed, it is not uncommon for contractors and architects alike to expect the arbitrator simply to 'split the difference' where the dispute is one over the valuation of variations. Not so. Though arbitration and litigation are apparently different in form and style, the parties should not confuse the difference between them with informality. Arbitration is a variant of formal legal proceedings.

Whether or not informality and an inquisitorial approach by the arbitrator would prove a more satisfactory approach is an open question, but that is not presently the case in practice. Except in certain limited circumstances the arbitrator will seldom test the parties' veracity by his own direct questioning and intervention. Neither is it usual for him to invite comment and/or response from either side. Present day arbitration, at least so far as the construction industry is concerned, is a far cry from that rather informal and inquisitorial approach. Like the judicial system, arbitration is adversarial albeit that it offers flexibility if the parties wish to take advantage of it.

Confidentiality is another important aspect of arbitration. Hearings are conducted in private not in an open court. Parties are free to

choose whether to represent themselves or whether to be represented and by whom. They need not be represented by a solicitor
and counsel in the traditional way. They may choose to represent
themselves or be represented by someone with expertise and
qualifications in one or more of the construction professions,
coupled with legal qualification and experience in the care and
conduct of such proceedings.

Employer and contractor are free to agree who should be
appointed, or should appoint, the arbitrator. They also have freedom to agree important matters such as the form and timetable of
the proceedings. This raises the possibility of a quicker procedure
than would otherwise be the case in litigation and matters such as
the venue for any future hearing can be arranged to suit the con-
venience of the parties and their witnesses.

11.8   *Joinder of parties*

Clause 41B.1.2 combined with rules 2.6, 2.7 and 2.8 of the CIMAR
are an attempt to introduce into the arbitration a type of third party
procedure similar to what is available in litigation, by enabling two
or more related arbitral proceedings to be heard by one arbitrator.
The intention of clause 41B.1.2 is no doubt to make provision so that
all parties will join in the arbitration if the dispute raises issues
which are substantially the same as or connected with issues raised
in a related sub-contract dispute about to be referred to arbitration.
The clause attempts to confer on the arbitrator powers which an
arbitrator would not otherwise have and aims to enable the same
arbitrator to determine all the disputes.

There is an obvious benefit to joining sub-contractors into any
main contract arbitration proceedings over related issues. To be
effective in relation to domestic sub-contracts a similar provision
would have to be inserted in all sub-contracts. In any event, the
clause does not, nor apparently does it seek to, provide the
machinery for joinder of the architect, quantity surveyor, mechanical engineer or other consultant.

11.9   *Procedure*

If the parties attempt to avoid the application of CIMAR, they
would have to make substantial amendment to article 7A along
with wholesale amendment to clause 41B. More to the point, such

amendment would be largely ineffective, because much of what now appears in CIMAR, and in the contract, merely reflects provisions of the Arbitration Act with which the parties and the arbitrator must comply. Any amendment to JCT 98 should not be undertaken lightly. Only after careful and expert consideration should such alterations be made and in the case of CIMAR it is difficult to think of any good reason why the parties contracting in England and under English law should want to avoid their use.

Arbitrations begun under contracts made using JCT 98 and subject to the law and jurisdiction of the English courts according to article 7A and clause 41B, must be conducted subject to and in accordance with the 1998 edition of CIMAR, current at the base date stated in the appendix. Notably, if any amendments have been made to those rules since that base date then the parties may jointly agree to instruct the arbitrator, in writing, to conduct the reference according to the more recent version (clause 41B.6). In addition to their express agreement to use CIMAR the parties also expressly agree that the provisions of the Arbitration Act 1996 shall apply too (clause 41B.5), irrespective of where the arbitration or any part of it will be conducted.

The new rules have much to commend them. They continue to offer the parties a choice of three broad procedures by which to conduct the proceedings, as follows.

11.9.1 Documents only procedure

Experience of disputes that have commonly arisen under earlier versions of this contract suggest that this will rarely be a viable option. Nevertheless, it is a much maligned and often ignored option that on occasion can offer real economies of time and cost. It is best suited to disputes capable of being dealt with in the absence of oral evidence and where the sums in issue are modest and do not warrant the time and associated additional expense of a hearing. Parties, either simultaneously or sequentially as the arbitrator directs, will serve on each other and on the arbitrator a written statement of case which, as a minimum, will include:

- An account of the relevant facts and opinions relied upon
- A statement of precisely what relief or remedy is sought.

If factual evidence of witnesses is to be relied on, then witness statements (or 'proofs'), signed or otherwise confirmed by the witnesses concerned, will also be included with the statement of case. Similarly, if the opinion of an expert or experts will be relied on those too will be given in writing, signed (or otherwise confirmed) and incorporated. There will be a right of reply and if any counterclaim is made, that too may be replied to before the arbitrator, if he wishes, puts questions or asks for further statements as he considers necessary or appropriate. Should he ultimately wish to do so the arbitrator may, and the rules provide that he can, set aside a day or less during which to question the parties and/or their witnesses. If that does happen then the parties will have an opportunity to comment on any additional information that may then have become known.

Given the type and size of issues most commonly suited to this type of procedure, more often than not the arbitrator will be in a position to reach his decision within a month or so of final exchanges and questioning.

11.9.2   Short hearing procedure

Although another unlikely option given the nature and complexity of disputes common to design and build contracts, this is a useful procedure which limits the time that the parties have within which to orally address the matters in dispute before the arbitrator. That time can, of course, be extended by mutual consent but without that agreement it is unlikely that more than one day will be allowed during which both parties will have a reasonable opportunity to be heard. Before then, each party will provide to the arbitrator and to each other a written statement of their claim, defence and counterclaim (if any), as the case may be.

Each such statement will be accompanied by all documents and any witness statements that it is proposed to rely on and if appropriate, the arbitrator will have the opportunity to inspect the subject matter of the dispute should he wish to do so. This is a procedure particularly well suited to issues which might readily be decided principally by such an inspection and is useful if the arbitrator can decide the issues and make his award within a short time-frame of around a month or so after considering the statements and having heard the parties.

In appropriate circumstances expert evidence can be presented by one, or more usually both, parties. It may not be necessary, particularly where the arbitrator has been chosen or appointed specifically with his own specialist knowledge and expertise in mind. In that case, parties can quite readily agree to allow the arbitrator to use his specialist expertise when reaching the decision. The use of independent expert evidence under the short procedure is almost actively discouraged under rule 7.5 of CIMAR which precludes any party which calls such expert evidence from recovering the costs of doing so, except if the arbitrator determines that such evidence was necessary for coming to his decision.

This short procedure with a hearing has many advantages over the often expensive procedure with a full hearing as discussed below. It is ideally suited to many common small and not unduly complex disputes and provides for a quick award with minimum delay. In practice there is nothing to prevent either the employer or contractor suggesting this procedure.

11.9.3   Full procedure

Where neither of the preceding shorter options is deemed satisfactory, or where, for example, there is significant disagreement over the essential facts or technical opinion evidence, rule 9 of CIMAR makes provision for the parties to conduct their respective cases in a similar manner to conventional High Court proceedings.

Like litigation, arbitration must offer finality in deciding not only simple but also the most complex of disputes. In such cases the contract and the rules must cater for the whole range of issues that might arise and must offer a workable framework around which the proceedings should be conducted. Consequently they must be capable of modification in the same way as they are in the High Court, so that they can be properly suited to the particular circumstances of any given disputes. For that reason the rules make clear that the arbitrator is free to permit or direct the parties at any stage to amend, expand, summarise or reproduce in some other format any statements of claim or defence so as to identify the matters essentially in dispute including preparing a list of matters in issue.

Parties operating under the full procedure will usually exchange formal statements comprising claim, defence and counterclaim (if any), reply to defence, defence to counterclaim and reply to defence to counterclaim. Each must be sufficiently particularised to enable the other party to answer each allegation made, and must as a minimum set out:

- The facts and matters of opinion which will be established by evidence and may include statements concerning any relevant point(s) of law
- Evidence, or reference to the evidence it is proposed will be presented, if this will assist in defining the issues
- A clear statement of the relief or remedies sought such as, for example, the specific monetary losses claimed set out in such a way as will enable the other party to answer or admit the claim made.

The arbitrator should give detailed directions concerning both the timing for service of the statements and all other procedures necessary in the period leading up to the hearing. Commonly the directions will include orders regarding the time within which either party may request further and better details of their opponent's case and the timing of the reply to any such request. Directions may also be given requiring the disclosure of any documents or other relevant material which is or has been in each party's possession. More likely than not, the parties will be required in advance of the hearing to exchange written statements setting out any evidence that may be relied upon from witnesses of fact. If expert evidence is also being relied on, the timing for preparation and exchange of written statements from experts will be the subject of a direction from the arbitrator.

11.10   *Appointing an arbitrator*

The parties need not wait until completion of the Works, or until determination or alleged determination of the contractor's employment under the contract, before proceeding to arbitration. Provided a 'dispute or difference as to the construction of this contract or any matter or thing of whatsoever nature arising thereunder or in connection therewith' exists between the parties, either party can begin arbitration proceedings. The first step in the procedure is for one party to write to the other requesting concurrence in the appointment of an arbitrator. In most cases, it will be the contractor who does so but there is no reason why the employer should not take the initiative. Whoever does so, the proceedings are formally begun when one or other party serves on the other a written notice, in the manner provided by rule 2.1 of CIMAR, 'identifying the dispute and requiring agreement to the appointment of an arbitrator'.

Article 7A is made subject to article 5. The effect is not immediately obvious, because article 5 simply allows either party to refer a dispute to adjudication [11.2]. Presumably, the draftsman intended any arbitration to wait until any adjudication was finished, but he did not say that and there appears to be nothing to prevent a party starting adjudication proceedings even though an arbitration on the same topic has already commenced. Certainly, adjudication can be started even though legal proceedings have been commenced: *Herschel Engineering Ltd* v. *Bream Properties Ltd* (2000).

When inviting agreement to the appointment of an arbitrator the party serving the notice should name at least one person that he proposes should act as arbitrator (CIMAR rule 2.2). It is good practice to suggest the names of up to three suitably qualified persons. Beyond the obvious requirements that nominees be competent, experienced and suitably qualified, they must also be independent. They must have no live connection with, or interest in, either of the parties. Nor should they have connections with any matter associated with the dispute. In choosing a suitable arbitrator it is essential for all concerned to have a proper understanding of the nature of the dispute and of the sums involved and so the parties must ensure that these are clearly appreciated from the outset. However appointed, the arbitrator must be independent, impartial, with no existing relationships with either the employer, contractor or anyone else involved or associated with the parties or the issue to be decided, and technically and/or legally qualified, as appropriate.

Wherever possible, it is sensible for the parties to make every effort to agree on a suitable candidate. Too often, requests for agreement on the appointment of an arbitrator are ignored entirely or rejected without thought. In such cases, deadlock is avoided by clause 41B.1.1 and by the provisions of rule 2.3 of CIMAR, both of which require that if the parties cannot agree on a suitable appointment within 14 days of a notice to concur or any agreed extension to that period, an arbitrator will be appointed by a third party. In a similar way to adjudication, when first completing the appendix the parties are given the choice of appointing bodies. All but one of those listed should be deleted. If no appropriate deletion is made or if no agreement is reached at the time of contracting, the task will fall, by default, to the President or a Vice-President of the Royal Institute of British Architects. He will make a suitable appointment after written application by either party.

Arbitral proceedings are begun in respect of 'a dispute' when one party serves on the other 'a written notice of arbitration identifying the dispute' and requiring him to agree to the appointment of an arbitrator. Notice of arbitration in connection with a particular dispute often provokes a counterclaim from the respondent. The words raise important questions about whether any such counterclaim can be brought within the jurisdiction of the original arbitration. It has long been the practice in construction disputes for respondents simply to raise their counterclaim formally at the time of serving their defence to the claim. That practice may now be obsolete, because it may be attacked by a claimant wishing to frustrate the respondents' attempts to automatically bring that counterclaim into the proceedings. Rule 3.6 of CIMAR makes clear that 'arbitral proceedings in respect of any other dispute are begun when notice of arbitration for that other dispute is served'.

Such tactics are more than merely point-scoring exercises. Although a claimant's insistence that the respondent serve a fresh notice in respect of his counterclaim is generally nothing more than a temporary hindrance, clause 41B.1.1 and rule 2.3 raise very real practical issues. Doubts over whether a counterclaim has properly been brought within the jurisdiction of the original arbitration may well have an effect on the existence and the extent to which either party gains protection from liability for costs where previous without prejudice offers of settlement have been made. Moreover, it is of considerable practical importance if it is only long after the initial arbitration has been commenced that the respondent or his representatives realise that a fresh notice, and hence new proceedings, are necessary in order to pursue a counterclaim. At its most extreme, the counterclaim might even then be time barred if the realisation dawns only after any contractual or statutory time limit for the commencement of proceedings has expired.

Notice to concur in connection with a previously undisclosed counterclaim may also be susceptible to attack as invalid on the basis that, at the time the notice was given no 'dispute' existed. Quite simply, it may be argued that no opportunity has been given for the subject matter of the counterclaim to be considered and rejected, or even possibly accepted, and so no 'dispute or difference' can yet be said to exist: *Hayter* v. *Nelson* (1990).

Subject to any statutory limitation, it seems clear that further notices to concur can be served either before an arbitrator is initially appointed (CIMAR rule 3.2) or after his appointment (CIMAR rule 3.2 and clause 41B.1.3 of JCT 98) and in the latter case CIMAR rule 3.3 will apply to the subject matter of that subsequent notice.

When determining which rules apply to the service of any further notice to concur, establishing the precise timing of the initial appointment of the arbitrator may well be important. If the arbitrator's appointment is made by agreement, it will take effect when he confirms his willingness to act, irrespective of whether by then his terms have been agreed. If, on the other hand, his appointment is the result of an application to the agreed appointing body, it becomes effective, whether or not terms have been agreed, when the appointment is made by the relevant body (clause 41B.1.1 and CIMAR rule 2.5). Arbitrators are professional men and charge professional fees. There is no fixed scale of charges for their services and individual's fees will depend on their experience, expertise and often on the complexity of the dispute. Respondents receiving a notice to concur should waste no time in taking proper expert advice on how best to respond. Claimants and counterclaimants would also be well advised to ensure strict compliance with the rules and should take proper advice both as to the timing and the content of any notice which they either intend to serve or have received.

The arbitrator will consider which of the procedures summarised above appears to him and to the parties to be most appropriate both in terms of not only the size and complexity of the particular dispute but also the procedure's suitability as a forum for the parties to put their own case and to answer their opponent's case. At the same time, the dispute must be kept in context. The arbitrator must have an eye to the format that will best avoid undue cost and delay and for even the most experienced arbitrator that is often a most difficult balancing act. Parties should, therefore, as soon as possible after his appointment, provide the arbitrator with an outline of their disputes and of the sums in issue along with an indication of which procedure they consider best suited to them.

After due consideration of both his own and the parties' views the arbitrator will generally arrange a meeting at which the parties or their representatives attend before him to decide upon which procedure shall apply. Specific time constraints may be imposed, or directions regarding the early hearing of certain questions of liability, or other preliminary matters, may be given. Although parties are always free to conduct their own case, if disputes have reached the stage of formal proceedings it is often better to hand over care and conduct of the proceedings to consultants with particular expertise in the care and conduct of arbitration proceedings.

11.11   *Powers of the arbitrator*

Under the 1950 and 1979 Arbitration Acts, arbitrators already had extremely wide powers. With the 1996 Act those powers in

important respects have grown significantly wider. The contract
(clause 41B.2) provides, subject to article 7A and to clause 30.9, that
the arbitrator has power to:

- Rectify the contract, i.e. to order the correction of errors if the
  contract fails to represent what the parties actually agreed. It
  must be shown that the parties were in complete agreement on
  the terms of the contract, but by an error wrote them down
  wrongly
- Direct such measurements and valuations as may in his opinion
  be desirable in order to determine the rights of the parties
- Ascertain and award any sum which ought to have been
  included in any certificate
- Open up, review and revise any certificate, opinion, decision,
  requirement or notice
- Determine all matters in dispute which shall be submitted to him
  in the same manner as if no such certificate, opinion, decision,
  requirement or notice had been given.

The last two powers are important. He can review the architect's
decisions and opinions and can, in effect, substitute his own opinion
for that previously formed by the architect. This is especially
important in the context of matters such as extensions of time and
the revaluation and payment of claims made by the contractor for
direct loss and/or expense.

Beyond specifically agreed limits to his powers provided for in
the contract, the arbitrator's statutory powers are considerable and
it should be realised that only in the most limited circumstances can
that power be revoked unless the parties have already agreed in
what circumstances that may be done, or otherwise act jointly in
writing to do so.

11.12   The legal procedure alternative

From the earliest RIBA contracts of the late 1800s through to the mid
1990s editions of their JCT counterparts, all of them in one form or
another incorporated provisions whereby the parties were required
to have their disputes determined by an arbitrator rather than by the
courts.

Unless the parties agreed to the contrary or there were exceptional reasons to do otherwise, they had no alternative; arbitration
was the only formal means available for breaking the deadlock if

they were unable to compromise. Significant increases in the value and complexity of construction claims have brought increases in the time, cost and expertise necessary to resolve such claims.

Following the decision in *Beaufort Developments (NI) Ltd* v. *Gilbert-Ash NI and Others* (1998), time will tell whether the previous long standing use of arbitration will continue as the usual method by which construction disputes are finally resolved. For some 14 years, between 1984 and 1998, the courts' power and jurisdiction to open up, review and revise architects' certificates and opinions given under JCT contracts was severely curtailed. Until *Beaufort*, the key judgment was *Northern Regional Health Authority* v. *Derek Crouch Construction Co* (1984). It had the effect of preventing the courts from opening up, reviewing or revising architects' opinions and certificates issued under the JCT family of contracts. Since construction disputes often raise questions about the correctness or otherwise of such certificates and opinions, the *Crouch* decision was perhaps the singular most important factor influencing parties in their decision whether to adopt arbitration or litigation as their preferred method of dispute resolution.

Crouch effectively gave the parties no realistic alternative. Arbitration was without doubt the most appropriate option. But, after *Crouch* was overturned and since the courts are no longer constrained in that way, it remains to be seen whether contracting parties will continue to favour arbitration.

Parties wishing to adopt litigation (clause 41C) in favour of arbitration must ensure that they complete the appendix correctly to reflect properly that intention. They must amend the standard form contract by deleting the reference to clause 41B in the appendix, for if they do not do so, by default, the arbitration agreement will take effect.

CHAPTER TWELVE
PRACTICAL COMPLETION, DEFECTS LIABILITY PERIOD, THE FINAL ACCOUNT

12.1   Practical completion: what is it?

The issue of the certificate of practical completion is dealt with in clause 17.1. The clause has undergone a succession of changes since the introduction of JCT 80. Before the architect can issue the certificate he must be satisfied about two, and possibly three, criteria:

- Practical completion of the Works must have been achieved in a physical sense
- The contractor must have complied sufficiently with clause 6A.4
- If work under clause 42 (performance specified work) has been carried out, the contractor must have complied with clause 5.9.

If the architect is satisfied, he must issue a certificate 'forthwith', i.e. as soon as is reasonable: *London Borough of Hillingdon* v. *Cutler* (1967). It is to be noted that the architect may only issue one certificate, not one when each of the criteria has been satisfied nor indeed when various parts of the Works are completed (unless of course the sectional completion supplement has been used).

After issue, the clause proceeds to state that practical completion is 'deemed', for all the purposes of the contract, to have taken place on the day stated in the certificate. The use of the word 'deemed' is particularly strange, because when a thing is deemed, it is tantamount to saying that parties will act in accordance with the deeming provision although knowing it to be false: *Re Cosslett (Contractors) Ltd, Clark, Administrator of Coslett (Contractors) Ltd in Administration* v. *Mid Glamorgan County Council* (1997). It is unclear why it is necessary to 'deem' practical completion when the contract clearly charges the architect with certification. The reference to 'all the purposes' simply means that whenever practical completion is stated in the contract to be the trigger or the closure of an event or situation, this certified date is the date referred to. That particular part of the clause is clearly otiose.

It is surprising that the term 'Practical Completion' is not adequately defined in the contract. It is true that it is listed in the definitions clause (1.3), but, on looking up the term, the enquirer is rather lamely referred back to clause 17.1 which does nothing to define it but simply, as has been seen, requires the architect to certify when it has occurred. The leading case is *Westminster Corporation* v. *J. Jarvis & Sons Ltd* (1970) and a most useful summary of the position was given by Judge John Newey in *H.W. Neville (Sunblest) Ltd* v. *William Press & Son Ltd* (1981). Essentially, practical completion has occurred when the Works are reasonably in accordance with the contract, when there are no visible defects and only minor things are left to be completed. In practice, the Works will never be completed, because there will always be a dab of paint or the twist of a screw between almost and entirely complete. Practical completion certainly does not mean entirely complete down to the last detail: *Emson Eastern Ltd (in receivership)* v. *EME Developments Ltd* (1991). Even when the architect is satisfied about this, he cannot certify if there are other outstanding criteria.

Clause 6A refers to the situation where the appendix states that the CDM Regulations apply. In practice, it is unlikely that they will not apply to projects carried out under this contract. Clause 6A.4 requires the contractor to provide and 'ensure' (i.e. guarantee that it will be done) that the sub-contractors provide the information reasonably required by the planning supervisor to prepare the health and safety file stipulated by regulations 14(d), 14(e) and 14(f) of the CDM Regulations. The information must be provided within the time 'reasonably required in writing' by the planning supervisor. If, which is unlikely, the contractor is not the principal contractor for the purposes of the Regulations, the contractor must supply the information to the principal contractor. The only person who can say when the second of the criteria for the issue of a certificate of practical completion is satisfied is the planning supervisor. It, therefore, appears that the architect must enquire whether the planning supervisor has received what he reasonably requires before deciding whether the contractor has 'complied sufficiently' with clause 6A.4. The position is anything but clear, because the insertion of the word 'sufficiently' to qualify 'complied' makes clear that the second criteria will be satisfied by something rather less than total compliance. Although the architect can do little but ask the planning supervisor if he has received what he requires, it seems that it is a decision for the architect whether the planning supervisor has 'reasonably' required it and also whether the contractor has 'sufficiently' complied.

Where performance specified work has been carried out, the third criterion requires the contractor to have complied with clause 5.9. Clause 5.9 is only relevant to performance specified work. It requires the contractor to supply as-built information concerning the work before practical completion. The information is expressly stated to include operational and maintenance information concerning the work and any installations forming a part of the work. This appears to refer to installations which may not themselves be performance specified work, but which form a part of it.

It is important to remember that practical completion is not defined anywhere as being the stage at which the employer decided to retake possession of the site.

12.2   Consequences of the certificate of practical completion

When the architect issues this certificate:

- The contractor ceases to be liable for liquidated damages
- The contractor's insurance obligations end
- The employer's right to deduct the full retention percentage ends
- The defects liability period begins to run
- The six months period within which the contractor must deliver all final account information to the architect or quantity surveyor begins
- The contractor's liability for frost damage ends.

12.3   Provision regarding partial possession

Where, with the consent of the contractor, the employer takes possession of any part of the Works before the certificate of practical completion of the whole has been issued, clause 18.1 provides that the architect must issue the contractor with a written statement which identifies the relevant part of the Works and gives the date of possession (the 'relevant date'). There are two things to note. The first is that the architect is issuing a statement and not a certificate. The architect is not giving his formal opinion that possession has taken place, he is stating it as a fact. The second thing is the reason for the first. It is not even the architect's statement, because he is said to give it 'on behalf of the employer'. The contractor may withhold his consent, but not unreasonably. In practice, a contractor could probably put forward very reasonable grounds for withholding consent in most cases, but he will seldom do that because of the advantage gained by allowing possession.

The taking of partial possession triggers several effects. The principal effect is that practical completion of the part is 'deemed' to have occurred for the purposes of clause 17.2, 17.3, 17.5 and 30.4.1 and the defects liability period for the part is 'deemed' to have commenced on the relevant date (clause 18.1.1). One effect of that is that the appropriate proportion of the retention money is released. The remainder of clause 18.1 wraps up the situation. Clause 18.1.2 provides for a separate certificate of making good of defects to be issued in respect of the part. Clause 18.1.3 deals with the insurance position and makes clear that the contractor's obligation under clause 22A or the employer's obligation under clause 22B or 22C.2 to insurance ceases on the relevant date. However, if the employer is insuring under clause 22C, the part taken into possession is included in the existing structures insurance under clause 22C.1. In a clause of mind numbing complexity (clause 18.1.4) the amount of liquidated damages payable by the contractor for the Works as a whole is reduced in accordance with the ratio between the contract sum and the contract sum less the value of the part taken into possession.

12.4   Defects liability period

The defects liability period is named in the appendix. The contractor is required to make good at his own cost any defects which appear within this period and which are due to workmanship or materials not in accordance with the contract or to frost damage occurring before practical completion. Clause 17.2 sets out the procedure. The architect must deliver to the contractor a schedule of defects in the form of an instruction no later than 14 days after the end of the defects liability period. The contractor is allowed a 'reasonable time' in which to make good the defects. What exactly constitutes a 'reasonable time' will depend on particular circumstances which will include the number of defects, their type and the difficulty of rectification. It is for the architect to decide when a reasonable time has expired and then to advise the employer and get his consent prior to taking decisive action. The clause expressly acknowledges that the architect may, with the employer's consent, instruct the contractor not to make good, followed by an 'appropriate deduction' from the contract sum.

The principle to be applied in deciding what an appropriate deduction may comprise is fairly straightforward. All defects are breaches of contract (despite the reference to them as simply 'temporary disconformities' by one judge: *P. & M. Kaye v. Hosier & Dickinson* (1972)). Therefore the defects liability clause in the contract amounts to a privilege and a right for a contractor who otherwise would have no right to re-enter the site after practical completion. Moreover, the employer would otherwise be entitled to employ other people to correct the defects. The employer is not limited by the clause. He could if he wished take action against the contractor for damages for breach of contract and the amount he could recover would be constrained only by the usual rules for recovery of damages for breach of contract. For example, he may be able to recover damages for loss of use of the premises during this period: *H.W. Neville (Sunblest) Ltd v. William Press* (1981).

In normal circumstances, if the employer decided that he did not want the contractor to rectify defects or if the architect failed to list some defects and subsequently the contractor refused to rectify them, the appropriate deduction from the contract sum would be simply what it would have cost the contractor to make good the defects: *William Tomkinson v. Parochial Church Council of St Michael* (1990). However, if the contractor refused to rectify defects which had been properly listed and notified to him at the correct time or if latent defects came to light after the end of the defects liability period, the architect would be entitled to issue an instruction to omit the rectification and get it done by others. The entire cost of the rectification, including professional fees, could then be deducted from the contract sum. In the case of a latent defect, the procedure would be for the employer to deduct the cost from any future payments or to recover the amount as a debt, rather like the procedure under clause 4.1.2: *Pearce and High v. Baxter*(1999).

In addition to the express provision for the serving of a schedule of defects on the contractor, the contract, in clause 17.3, provides the architect with a very useful additional power. Whenever he considers it necessary, the architect may issue instructions requiring the contractor to make good specific defects. The architect is given a very broad discretion. The criterion is simply his opinion that the instruction is necessary. There appears to be nothing to prevent the architect from issuing such instructions one after the other as defects are noticed. The only proviso is that he may not issue a clause 17.3 instruction after the expiry of 14 days after the end of the defects liability period or the architect has issued his defects schedule under clause 17.2, whichever is the earlier.

When the contractor has made good all the defects which have been notified to him under clauses 17.2 and 17.3, the architect must issue a certificate of completion of making good defects (clause 17.4). Completion of making good defects is 'deemed' to have taken place on the day stated in the certificate. It is suggested that the use of the word 'deemed', which is otherwise unnecessary, can only mean that, on the issue of the certificate, the parties agree to treat all the notified defects as being made good, even if the architect has overlooked some defects. The result appears to be that the employer cannot subsequently require the contractor to rectify such over-looked defects. The same consideration does not apply to latent defects. Many architects become unnecessarily concerned about issuing a certificate of completion of making good defects if further defects have appeared after the issue of the schedule. There is no need for concern, because the certificate merely refers to notified defects and not to subsequent ones. However, there is an important consequence which certainly does have a bearing on the amount of money available to deal with defects.

12.5   Release of retention monies

The contractor is entitled to the release of one half of the retention money in the next interim certificate after practical completion [5.6].

The balance is to be released in the next interim certificate after the issue of the certificate of completion of making good defects. Clause 30.1.1.2 states that the employer is entitled to exercise any right under the contract of deduction from monies due to the con-tractor in any interim certificate whether or not that interim certi-ficate includes retention monies. This is subject to qualifications about the deduction from interim certificates of monies paid direct to nominated sub-contractors – see clause 35.13.5.3.2 [4.9].

12.6   Penultimate certificate

It is sometimes overlooked that, under clause 30.7, the architect must issue an interim certificate to include the finally adjusted sub-contract sums for all the nominated sub-contracts. He is to do this 'so soon as is practicable'; in any event there must be a minimum of 28 days between the issue of this certificate and the final certificate. As soon as the architect is in a position to issue it, the interim certificate must be issued even if the previous certificate was issued less than a month previously.

12.7　Final account

The contract does not refer to a 'final account' anywhere although this is the name which everyone gives to the statement drawn up at the end of a contract to show how much is owed, usually, by the employer to the contractor and how that figure is derived. Instead the contract refers to a statement of adjustments to the contract sum and ascertainment under clauses 26.1, 26.4.1 and 34.3. However, what they clearly mean is the final account.

Since JCT 98 is a lump sum contract, the final account is prepared by taking the contract sum and making deductions from it and then making certain additions. For this purpose, the valuations and assessments made for interim certificates are ignored. Clause 30.6.1.1 requires the contractor to provide the architect or, if the architect so instructs, the quantity surveyor with everything necessary for the purpose of calculating the final account. He is given six months after the date of practical completion to do this. In practice, the contractor usually does this in the form of a fully documented final account of his own.

It will be a poor quantity surveyor who has not been constantly calculating the predicted final account as the project progresses. It will be necessary to keep both the architect and the employer updated about the financial standing of the contract. Therefore, the six month period is the contractor's last opportunity to bring to the attention of the quantity surveyor any items which might have been overlooked. It is common for contractors to be very slow in complying with this requirement and late delivery of documents frequently delays the completion of the quantity surveyor's final account which he must complete and send to the contractor (together with appropriate extracts for each nominated subcontractor) within three months of receipt of the contractor's information. It has been made clear that neither the supply of documents by the contractor nor the issue of the quantity surveyor's calculation of the final account is a pre-condition to the issue of the final certificate: *Penwith District Council* v. *P. Developments Ltd* (1999).

Clause 14.2 expressly prohibits the adjustment of the contract sum other than as set out in the contract. It also makes clear that, except for the provisions of clause 2.2.2.2 **[1.9]**, errors in the computation of the contract sum, i.e. the sum stated in article 2, are deemed accepted. In other words, the final calculation of the amount due is not an opportunity to open up the original article 2 figure.

Clause 30.6.2 sets out the way in which the final account is to be calculated. It is divided into three broad sections. The first deals with 'adjustment' which may take the form of adding or deducting value:

- Clause 13.4.1.1 valuations agreed between employer and contractor
- Clause 13A quotations and subsequent variations
- Accepted price statements.

The second deals with deductions which must be made:

- Prime cost sums and amounts, clause 35.1 named sub-contractors and amounts paid by the employer in respect of defective nominated sub-contract work
- Provisional sums and approximate quantities
- Value of omitted items
- Amounts which the employer is entitled to deduct under clauses 7, 8.4.2, 17.2, 17.3, 38, 39 or 40
- Any other amount the contract states must be deducted from the contract sum. This appears to be a 'just in case we have forgotten anything' clause.

The third deals with additions which must be made:

- The nominated sub-contract final accounts
- The appropriate sum following acceptance of a clause 35.2 tender
- Amounts payable by the employer in respect of nominated suppliers
- Contractor's profit at appropriate rates on the previous three items
- Amounts paid by the contractor under clauses 6.2, 8.3, 9.2 and 21.2.3
- Valuations under clause 13.5, except omissions
- Valuation of all provisional sum instructions and all approximate quantity work
- Ascertainment of loss and/or expense under clauses 26.1 or 34.3
- Amounts paid by the contractor unless under clauses 22B or 22C which should be added to the contract sum
- Amounts payable to the contractor under clauses 38, 39 or 40

- Any other amount the contract states must be added to the contract sum. This appears to be another 'just in case we have forgotten anything' clause
- Any amount accepted under the priced statement (clause 13.4.1.2) provisions instead of being ascertained under clause 26.1.

Clause Number Index to Text

Table of Cases

Index